THE COLORADO RIVER

PORTRAITS OF AMERICA

THE
COLORADO
RIVER

VIRGINIA HOPKINS

CHARTWELL
BOOKS, INC.

A QUINTET BOOK

Published by Chartwell Books Inc.,
A Division of Book Sales Inc.,
110 Enterprise Avenue,
Secaucus, New Jersey 07094

ISBN 0-89009-883-2

This book was designed and produced by
Quintet Publishing Limited
6 Blundell Street, London N7
in association with Footnote
Productions Limited

Art Director Peter Bridgewater
Editor Sheila Rosenzweig
Photographer Trevor Wood

Typeset in Great Britain by
Leaper & Gard Limited, Bristol
Colour origination in Hong Kong by
Hong Kong Graphic Arts Company Limited,
Hong Kong
Printed in Hong Kong by Leefung-Asco
Printers Limited

CONTENTS

PART I

THE
UPPER COLORADO

INTRODUCTION

he Colorado River isn't the longest or the widest river; it doesn't carry the most water and, surprisingly, it isn't even the most dammed; but it is the most controlled and controversial river in the world. Water is scarce in the West, and a normally taciturn westerner can be provoked to rage by those who want to take away his water. Water rights are jealously guarded by even the smallest ranch, and real estate is far more valuable with water rights than without. A scarcity of water is the price human beings pay for living in the desert, and as more and more people move to the warm, dry, sunny climate of the Southwest, the problem grows more and more serious.

Those living on the Western Slope of the Rockies in Colorado grumble about their water being diverted to the dry Eastern Slope. Arizonans grumble about water being diverted to California, and the Indians grumble about water going to swimming pools and toilets when it could be used for irrigation. (According to Colorado River expert Philip Fradkin, more Colorado River water goes to water the fields that grow the food that feeds the cattle that make the beef that Americans consume in such huge quantities than to any other single use.) Mexico grumbles because by the time its share of Colorado River water gets over the border it is so saline that it kills their crops. River runners and conservationists say the dam builders are ruining vast stretches of pristine canyon country with their reservoirs, while the dam builders point out that the same people are using that stored water every day. The Upper Basin states in the Colorado River Basin point to the greediness of the Lower Basin states in water allocation, then turn around and offer to sell what they aren't using.

Not only does everyone disagree on the fate of the Colorado River, but nobody is coming up with solutions, either. Major John Wesley Powell, the one-armed Civil War veteran who was the first to descend the Grand Canyon, predicted a shortage of water in the West almost a century ago. Every drop of the Colorado River is allocated, every mile below the Grand Canyon is as tame and controlled as an irrigation ditch, which, in effect, is what it had become at that point. But the river has been there for eons, and what humans do to it is a blink of the eye in geologic time. Chances are the Colorado River will be wending its way from the Rockies to the Gulf of California long after people go the way of the dinosaurs.

The area drained by the Colorado River equals nearly a twelfth of the continental United States, a total of 244,000 square miles. From the headwaters of the Green River (geologically the Colorado River's true continuation) its length is 1,700 miles; from its headwaters in Colorado, it is 1,400 miles long and descends 14,000 feet.

The Colorado does possess a few superlatives—it carries the most silt of any river in America, it has the warmest water, and it has the highest evaporation rate. The Colorado is not polluted by industrial and urban waste, but it is polluted by the saline runoff from irrigation. The quality and quantity of its water ranges from clear green glacial streams in the Wind River Range to the massive rapids of Lava Falls in the Grand Canyon to the sluggish and silty red of the lower river.

Members of Major Powell's **historic expedition** pause on the dizzy heights of the Grand Canyon's inner gorge.

Previous page The Flaming Gorge Recreation Area is the first large resevoir on the Green River.

Major John Wesley Powell, the first man known to travel through the Grand Canyon, was awed by the sheer walls of Marble Canyon **opposite**.

THE
UPPER COLORADO

Early explorers called the Green River the continuation of the Colorado River, and called the Upper Colorado the Grand River. Technically, both rivers are upper forks of the Colorado. The name of the Grand was switched in the early 1900s for political reasons: a Colorado senator decided it would be good public relations to have his state's river called the Colorado, and so it was.

The Headwaters

The Upper Colorado River doesn't come to life in a dramatic fashion, but from countless small snowmelt streams that trickle down from the peaks and glaciers of the Never Summer Mountains. However, the setting of the Colorado's headwaters is dramatic—Rocky Mountains National Park high in the north central Rocky Mountains of the state of Colorado. The stream that begins near La Poudre Pass continues past the site of Lulu City, a boom-and-bust mining town, and heads out the Kawuneeche Valley, down to Grand Lake, and from there to the Grand Valley below. The Central Colorado Rockies also drain the two forks of the Platte river, the Arkansas River to the east, the Rio Grande to the south, and the San Juan to the southwest. There are 52 peaks over 14,000 feet in Colorado, and 1,500 over 10,000 feet, all collecting the snow and rain which create these rivers.

The massive forces that shaped the present day Rocky Mountains began at least 100 million years ago when the entire region was covered by a sea that stretched from the Arctic to the Gulf of California. In the Rockies, the climate was subtropical, and giant reptiles ruled the land. The sandstone and shale that forms much of the park often contains the fossils of these creatures and other long-extinct plant and animal remains. During the next geological epoch, forces came from underneath the sea that lifted and buckled the land, creating the beginnings of the front range of the Rockies. As the eons passed, ero-sion, volcanoes and more erosion created a rolling upland during the Mesozoic age. During the Cenozoic age, about 50 million years ago, the earth buckled again, creating the raw form of the uplifted range of today. Since then many factors have been at work, including wind and water erosion, heat and cold, the carving action of glaciers (the remnants of which exist everywhere in the Rockies) and more volcanic eruptions.

Grand Lake is the largest natural lake in Colorado, and one of the highest yacht anchorages in the world. The Ute and Arapaho Indians who lived in the area called in Spirit Lake. The name came about, legend has it, one winter when the lake froze over completely except for a hole in the middle. There were many buffalo tracks in the snow on the lake, and the tracks of one huge buffalo in particular seemed to originate in the center of the lake and return there. To the Indians, this meant that the lake was the home of a god, and in recognition of this, it was named Spirit Lake. Though the Utes and Arapaho were the last Indian tribes to live in the area before the white men took over, there had been various groups of Indians living around Grand Lake for at least 2,000 years.

Explorers and Miners

The first white man to take an active interest in exploring the length of the Colorado River was a trapper named James Ohio Pattie. Young Pattie set out from the confluence of the Gila and Colorado in 1826, and came fairly near the headwaters of the Colorado in the Rockies, though he and his party were apparently hampered by deep snow.

The first white men to actually settle in the lush and spectacularly beautiful Estes Park area were Joel Estes and his son Milton, who built a cabin in the grassy meadows at the bottom of the valley in 1867. A few years later, a famous big-game hunter, the Earl of Dunraven, visited what would one day be Rocky Mountain National Park and was so enchanted

Opposite Through most of the Grand Canyon, the Colorado River is a vivid blue, in contrast to the red rock. Elsewhere along its course, the river is often red with silt.

The Stanley Hotel in Estes Park **above** is a grand old luxury hotel with fantastic views of the Rockies.

The Maroon Bells and the lake below them **below** provide one of the best loved vistas in all of Colorado.

with the area that he bought Estes' property. Much of the beauty of the area was preserved by the Earl, who brought many dignitaries and celebrities to see it.

Bouyant rubber rafts have made river running a safe and exhilarating experience, but back when the West was first being explored by white men, their roughly-hewn wooden boats weren't up to the rigors of most mountain rapids. About the same time Major Powell was preparing to descend the Grand Canyon, a young and foolish man named Samuel Adams sent the following message to Congress:

A singular fatality has from the first been connected with this history of the Colorado river. ... The letters written respecting it, and the continued effort made by a formidable coalition of corporations (for selfish purposes) to crush out the individual enterprise of proving its national importance, have all been of that revolting character as to do her the most flagrant injustice. The Colorado must be, emphatically, to the Pacific coast what the Mississippi is to the Atlantic. The building timber and ties for the construction of the railroad crossing and continent (now completed) have been carried upon one of her tributaries; and one of her grand purposes will not be completed until the material for the construction of the southern line is borne upon her surface.

To prove that everyone else was exaggerating the mightiness of the Colorado River, and that it could be used as a major highway for transportation from beginning to end, Adams decided to travel from its headwaters to the Gulf of California in wooden rowboats. A few years earlier, he had made a few unsuccessful tries at ascending the river from the Gulf of California by steamboat—now he would conquer it from the top. He had rough wooden rowboats built in Breckenridge, Colorado, and began his journey by heading down the Blue River with eight men and four boats. By the time he reached Gore Canyon, he had lost all of his boats and most of his supplies to the rapids. However, Adams' self-righteousness was only given fuel by his notion that nothing in the Grand Canyon could approach the rapids and height of Gore Canyon! He continued on crude rafts, and after a trip of perhaps 200 miles, Adams had lost a total of four boats,

Right Plodding, patient and sure-footed mules have been carrying people up and down the cliffs of the Grand Canyon since the early 1900s.

four rafts and eight men. He and the remaining three finally gave up somewhere in the Grand Valley, about the time that Powell and his men were stumbling out of the Grand Canyon, having covered close to 2,000 miles and surviving some of the most treacherous rapids in the world.

In 1868, Major Powell and a party of men made the first recorded ascent of Long's Peak, and on that trip Powell also saw Grand Lake, which lies in back of Long's Lake. This was his first glimpse of the headwaters of the river that was to occupy much of his life and make him famous.

Within ten years of Major Powell's ascent of Long's Peak, silver and gold had been discovered in Colorado, and a man named Joe Shipler staked a claim on Shipler Mountain, below La Poudre Pass. He was among the first miners to arrive and he never left, even though his mine never produced much ore. He had obviously fallen in love with the mountains, as so many have since, and wrote, "Climate very mild, any quantity of game in the parks: have caught as many as 583 trout in one day. Grazing is excellent ... grass grows knee high along the streams. Heavy timber in the mountains."

Others searching for gold followed Shipler. Further up the Colorado River, Lulu City was established in 1880. Hundreds of miners staked their claims and filled the air with the sound of blasting. The town had mail delivery, a stage line, a clothing store, a barber shop, a hardware store, and a couple of grocery and liquor stores. The place was booming, with a population of 500. Just a few years later, it became apparent there weren't going to be any great silver strikes in Lulu City, and it was too expensive to get the low-grade ore out of the mountains to a refinery without a railroad. By early 1884, the city was practically deserted, and *The Colorado Miner* reported that, "Much anxiety is felt for the safety of the mines at Lulu. Since the departure there of Judge Godsmark and some of the old timers, the bears and mountain lions have taken possession of the boys' houses and old, discarded overalls and gumboots, and are running a municipal government of their own, to wit; using all their efforts to restore Lulu to its primeval status."

And indeed, Lulu City is no more and can hardly even be called a ghost town, but for the remains of a few old stone foundations and fireplaces, the rusting carcasses of discarded mining machinery and a few old roads.

Just over a century ago the Grand Canyon was an unknown, unexplored chasm in the earth. Today the number of people allowed to run it in boats **left** has to be limited by the National Park Service.

Green patches of civilization stand out in the barren red rock country **center** that surrounds the lake.

The **black bear**, the most widespread of all North American bears, roams in mountain forests. Despite its name it is often a light brown.

A float trip down the Colorado
River reveals the formations of Grand Canyon
from a different perspective.

Snowmelt from the Never Summer
Range in Rocky Mountain National Park forms
the beginnings of the Colorado
River.

Aspen trees **left** turn the Rockies into a golden paradise in the fall, and provide a habitat for deer and elk.

Towering above Nymph Lake in Rocky Mountain National Park, Long's Peak **right** is one of the highest in the Rockies, and a favorite challenge of expert rock climbers.

Rocky Mountain National Park

The history of the Colorado River is also the history of water diversion in the United States. One of the first major diversion ditches was built in 1890 close to the headwaters of the Colorado. Today the Grand Ditch is visible as a long gash across the Never Summer Mountains, and, nearly 100 years later, carries about 30,000 acre-feet of water to the dry and heavily populated cities of the Eastern Slope of the Rockies. An acre-foot would be about the equivalent of a football field covered in ten inches of water. Most of the major streams of the Western Slope of the Rockies are diverted at some point to the thirsty Eastern Slope. Massive tunnels drilled right through the Continental Divide, dozens of dams, ditches and reservoirs all bring water from the wet side of the mountains to the dry side of the mountains. So from its very beginnings, the Colorado is diverted, and in spite of these unromantic and unsightly origins, it still remains one of the most awesome rivers on the planet.

Since its christening as a National Park in 1915, Rocky Mountain National Park has been spared much of the unsightliness of man's attempts to change the course of nature. This designation came about largely through the dogged efforts of naturalist Enos Mills, who is known as the "father" of the park. Today, over

19

The delicate and endangered columbine **right** is the Colorado state flower—enjoy but do not pick!

Summer wildflowers **opposite** turn the mountainsides of Rocky Mountain National Park into a blanket of color.

The **quaking aspen**, which grows throughout the Rocky Mountain range, takes its name from the near-ceaseless fluttering of its broad leaves.

Overleaf The fall colors enhance the majesty of the Rocky Mountains.

20

three million people visit Rocky Mountain National Park every year.

"This is a beautiful world, and all who go out under the open sky will feel the gentle, kindly influences of nature and hear her good tidings." Those are the words of Enos Mills, and visitors to Rocky Mountain National Park tend to take them to heart. Estes Park is now a busy little city, hosting hikers, river runners and sightseers in the summer, and in winter, skiers who come to challenge the slopes of the Hidden Valley ski area. In the fall, when the aspen trees blanketing the mountains turn a brilliant yellow, thousands of people drive the Trail Ridge Circle to "see the color," or head south from Poudre Lake to follow the North Fork of the Colorado to Grand Lake.

At dusk and dawn in the autumn, strange cries ring out in these forests—they are the bugling of male elk in rutting battles, which can be observed by those with patience. The bighorn sheep that live in the park are shy, and usually stick to the rockiest and most inaccessible crags, but with the help of a local guide, it is often possible to spot a few. July and August bring out thousands of tiny and delicate wildflowers which bloom on the alpine tundra.

Lower in the meadows and aspen forests, Colorado's state flower, the spectacular columbine, blooms.

Down through the Grand Valley

As the Colorado River thunders down the Western Slope of the Rockies, it is swelled by dozens of small streams. When it finally emerges from the Rockies, it is joined by the Eagle River, another good-sized mountain torrent. It then plunges abruptly into Glenwood Canyon, a magnificent stretch of vertical cliffs carved out by the river, and paralleled by a railroad and a highway. After its 20 or so miles through Glenwood Canyon, the river enters the Grand Valley, where it meanders until it gets to Utah.

About halfway through Glenwood Canyon, a short but steep trail heads up to Hanging

Fishing for native cutthroat trout **left** is a favorite pastime in the Rocky Mountains.

In the bucolic Grand Valley **right**, the Colorado River gives no clue of its icy mountain origins, or of the powerful rapids yet to come.

Lake, a pristine gem carved out by glaciers and fed by Bridal Veil Falls. A few miles further down the river is the Shoshone Power Plant, which supplies electricity to the Denver area. This is where kayakers and the more intrepid rafters begin local Colorado River trips. In the spring, when the water is high, there is almost always a good show to be had here as kayakers negotiate the rapids below Shoshone.

At the head of the Grand Valley is Glenwood Springs, a quiet little city known best for its giant hot springs pool and vapor caves, fueled by underground cauldrons. Chances are that people have been soothing away their aches, pains and worries in these pools for many centuries. The pools are a special treat in the winter, when the mist from their heat rises off the water in great billows.

Where the town of Glenwood Springs ends to the north, the Flat Tops rise to 10,000 feet and level out to a broad mesa of pine forests, lakes and meadows filled with wildlife that attracts fishermen and hunters from all over the world.

On the south side of Glenwood Springs, the Roaring Fork River joins the Colorado. Heading up the Roaring Fork towards Aspen, Mount Sopris rises dramatically out of the valley to the west. This impressive mountain is snow-capped year-round and is a favorite challenge of cross-country skiers. The Crystal River passes in the shadow of Mount Sopris to join the Roaring Fork at Carbondale, a small town with growing pains created by the influx of people hoping to work in the shale oil industry to the west.

Broad-winged hawks migrate to Colorado's mixed woodlands from their breeding grounds in Alberta.

Rain from afternoon thundershowers
on the rim may never reach the canyon floor
and the Colorado River, but
may send torrents of muddy water down side
canyons.

At the town of Basalt, the Fryingpan River joins the Roaring Fork. Though the Fryingpan has been harnessed near its source by Reudi Dam and Reservoir, it still offers some of the best fly-fishing-only water in the United States.

The last town on the Roaring Fork River is Aspen, one of the loveliest mountain towns in Colorado, and one of the most famous. Though Aspen has a reputation as an enclave of the jet set and as the site of four superb ski mountains, it also offers a myriad of cultural activities in the summer, from the top-notch Ballet Aspen to the Aspen Music Festival and the Aspen Film Festival. Like most towns in the Rockies, Aspen began as a rough-and-tumble mining camp, which at its height had a larger population than it does now. These days, Aspen retains its Victorian character and charm, thanks to strict zoning regulations that encourage renovation of the splendid old gingerbread mansions and forbid neon signs and tall buildings.

Miners and Mountains

In the early 1800s, the land west of the Mississippi River was largely unknown territory. Lewis and Clark made their famous trek across the country, returning with much information and many tales, but they had taken a northern route, avoiding the formidable barrier of the southern Rockies, the mazes of canyons south of Salt Lake City, and the inhospitable deserts of the Southwest. Explorers, traders and trappers made a few inroads into the wilderness, but until gold was discovered in California the West belonged to the Indians and the Spanish.

The first gold rushes in California inspired hundreds of thousands of people to head west to make their fortunes. Many of them went by boat, either around Cape Horn or to Panama, where they crossed the jungle on foot or horseback, then continued north by boat again to San Francisco. The wagon trains came in a trickle at first, establishing routes that thousands would later follow. These great migrations and a desire by the United States Government to find out whether the land was worth taking from Mexico inspired exploration of places like the Rockies.

Early topographers, such as Captain Zebulon Pike and Major Long, sent reports back to Washington that described the land up to the foothills of the Rockies as a desert. For decades, the Rockies of Colorado belonged to the fur trappers, the legendary mountain men who were later responsible for guiding many a wagon train and establishing the trails on which they forged west. It wasn't until gold and silver were discovered in the Rockies in the mid-1800s that people began settling on the Front Range, the Eastern Slope of the Rockies.

The first silver strike was located at Cherry Creek, an area from which the whole present-day metropolis of the Front Range would eventually spread. That strike brought in a few hundred people, but it was the rich lode found at Clear Creek that brought 15,000 people to the area in a matter of months, and over 100,000 within a year. Over the next 20 years, other major veins were found in Leadville, Cripple Creek, South Park, Aspen and dozens of other spots, and towns grew up around them. Many people were made millionaires overnight, becoming historical figures and Colorado legends. H.A.W. Tabor ran a grocery store in Leadville, and had a policy of grubstaking the miners who came through with claims but no money to buy supplies. When a couple of German prospectors gave Tabor a third of the Little Pittsburgh Mine in exchange for a grubstake, Tabor had it made—he eventually raked in 20 million dollars. Tabor lived an ostentatious life, building mansions for himself and his mistress, Baby Doe, erecting public buildings in Leadville and Denver, and throwing his money into whatever opportunities came his way. He frittered away his fortune and died poor. Baby Doe, who had become his wife, returned to Leadville. Tabor's dying words to her had been, "Whatever happens, hold on to the Matchless [mine]. It will give you back all that I have lost." Though Baby Doe guarded the Matchless with a shotgun for years, she died of cold and hunger in a shack in Leadville.

Another magnate built a castle at Redstone, Colorado, and tried to operate a company town in the style of a feudal baron. That failed,

Privately struck **gold coins**, called 'Pioneer Gold', were commonplace at the height of the Gold Rush in the 1850s.

Vail **opposite above** is a ski resort that grew up almost overnight, and though it lacks the charm of older Colorado towns, its slopes offer some of the best skiing in the country.

but tours are still given of the Redstone Castle, which is said to be haunted.

The fates of the mining town were as varied as their histories. Some, like Lulu city, disappeared altogether, and others, like Cripple Creek, became ghost towns, museums of past with wooden buildings preserved by the dry mountain air. Aspen might have faded into oblivion but for a new sport that quickly caught on—skiing. Leadville is still a mining town, barely hanging on, largely supported by the Climax mine operated by AMAX, which produces three-fifths of the world's molybdenum. Tabor's Vendome Hotel and Opera House can still be seen in Leadville. Cherry Creek was absorbed as a suburb of Denver. By the early 1900s, silver had been replaced by gold as the money standard, and Colorado

sank into relative oblivion for decades. The state's main industry was cattle ranching, and Denver became a sprawling cow town.

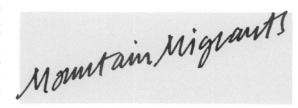

It wasn't until the 1960s, when people started fleeing American cities in droves, that the fresh mountain air and wide-open spaces of Colorado inspired another population boom. The Front Range metropolis of Colorado Springs, Denver, Boulder, Fort Collins and a

Leadville **left** has been a mining town for over a century — first silver, now molybdenum. It's one of the highest towns in the Rockies.

A raft trip on the Colorado **below** can be wild and wet or slow and lazy, and can last anywhere from a hour to weeks.

host of smaller cities, where 80 percent of the state's population lives, has a growth rate that is triple the national average. Manufacturing and industry account for the largest chunk of income in the state, followed by livestock and agriculture, mining and tourism. Within the manufacturing business, the high technology, electronics and computer industries have created another "Silicon Valley" on the Front Range. Other large producers include steel, film, rubber products, construction materials, commercial buses and of course beer, led by Coors, which is located in Golden Colorado. Agricultural products include beef, wheat, corn, hay, sheep, potatoes, pigs, beans, turkeys, barley, sugar beets and sorghum. Colorado mines produce petroleum, uranium, tin, molybdenum, gold, silver, coal and more than 250 minerals.

With more than 30 ski areas, most with excellent conditions throughout the winter, Colorado ranks as the number one skiing state in the nation. Resort towns such as Aspen, Vail, Crested Butte, Steamboat Springs and Estes Park have economies almost completely dependent on tourism, and a high percentage of young and physically fit residents. There are also many transients in these towns. Being a ski bum is still a popular way for college students to spend a year or semester off; others come and go with the ski season as a way of life. A ski resort can easily quadruple its permanent population with tourists and transients during the winter. Every resort has its jet setters and celebrities who buy homes or condominiums and dazzle the town for a few

Coors brewery, famous for its beers and its advertisements for generations, remains one of Colorado's leading industries.

Colorado National Monument **above** near Grand Junction is the beginning of red rock and canyon country on the Colorado River.

Though Great Sand Dunes National Monument **right** is far removed from the Colorado River, it's worth a detour.

weeks or months every year. On major holidays, Aspen's airport is crowded with private planes and jets, and the streets are crowded with people hiding behind sunglasses and fur coats. Jobs at the ski areas as ski patrol members, teachers and lift operators are highly coveted among resort residents. Frequently, young couples who have made good financially on the east or west coats flee the cities and open small businesses in these resorts. Many who decide to make the mountains home and like the resort lifestyle make a living catering to skiers in the winter and also to summer tourists who come for whitewater raft trips, hiking, rock climbing, fishing, horseback riding or just some clean, crisp mountain air and sunny days. Golf and tennis are particularly popular at the high-altitude resorts because balls travel farther in the lighter air. All of the major resorts have

catered to these tastes with championship golf courses and dozens of tennis courts.

Tourists pour into Colorado by the thousands to see Mesa Verde National Park, a huge Pueblo Indian ruin in southwestern Colorado; Rocky Mountain National Park; Great Sand Dunes National Monument; Colorado National Monument, and Dinosaur National Monument. Most towns with "Springs" in their name do have hot springs to soak away the aches and pains of a day out-doors: Glenwood Springs, Idaho Springs, Pagosa Springs, Eldorado Springs, Hot Sulphur Springs, Steamboat Springs and many others. Dozens of dude ranches take people into the high country on horseback for a "back to nature" experience that is generally quite luxurious. Big game hunting brings in hunters from all over the country in the fall, and the trout streams and lakes bring in fishermen.

The male of the **vermilion flycatcher**, found near desert water, is at once recognizable by its flaming red crown, throat and breast.

Cliff Palace **left** at Mesa Verde National Park is one of the largest and best preserved pueblo ruins.

Overleaf Independence Monument at Colorado National Monument was formed by erosion, which continues to wear away at this part of the Uncompahgre Highland. **31**

THE GRAND VALLEY

As the Colorado River winds its leisurely way west through the Grand Valley, its water is used for irrigation, and what drains back into the river begins an increase in the water's salinity, which continues to rise all the way to the Mexican border. Much of the valley west of Glenwood Springs is fertile bottom land with a good growing season. Cattle graze in the fields that line the river, and alfalfa and other livestock food crops are irrigated by the Colorado and the streams that come down from the mesas to the north and south.

The desert-haunting **roadrunner**, a species of cuckoo, makes its unique clattering sound by rubbing its mandibles together.

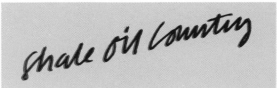

Shale oil Country

The towns of Rifle, Silt and Parachute were quiet little backwaters until a few years ago when they became the scene of frenzied construction, speculative land-buying and development. United States oil companies have·known for a long time that the shale in the western Grand Valley was rich with oil— the only thing stopping them from extracting it was the lack of a cost-effective technology. Spurred on by the oil crisis and billions of dollars in feudal loan guarantees, a number of oil companies decided to go ahead with oil-shale development in the late 1970s. In the small, economically impoverished towns nearby, huge mobile-home parks appeared overnight, new sewage systems, roads and schools were built, and thousands of people moved into the area from all over the country. To the south, up on Battlement Mesa, a whole new town was built, complete with condos, housing tracts, churches, schools, shopping malls and even a plush athletic club. Within a few years, before the town of Battlement Mesa was even completed, it became clear that once again, on a large-scale basis, the oil-shale industry was going to have to be shelved for lack of cost-effective technology. Most of the oil companies moved out, leaving a dazed populace and a brand-new ghost town at Battlement Mesa.

One of the key elements in the failure to develop the oil shale was water, a scarce and valuable commodity in that part of Colorado, even with the Colorado River flowing through the valley. Though the oil companies had been buying up land for decades, and with it precious water rights, the amount of water needed to process the shale and extract the oil was hundreds of times the amount available.

As one enters Grand Junction, fruit stands being to proliferate. To the south and east of Grand Junction, separated from the Colorado River by a series of mesas, is a lush valley filled with orchards, where some of the best peaches in the United States are grown. Fruitvale, Orchard City, Delta, Hotchkiss and Paonia are all part of this fruit bonanza which also produces apples, cherries, pears and vegetables. With the recent introduction of vineyards, the valley may one day be known for its wines as well. Nourished by rich soil and water from the surrounding mountains, these orchards also owe their existence to weather patterns that inflict the worst of the cold and snow on the mountains to the east, and the worst of the heat and drought on the arid lands to west.

Grand Junction

The Colorado River runs right through the middle of downtown Grand Junction, forming a centerpiece for the largest city on the Western slopes of Colorado. Grand Junction also has an unusual climate because it is located in a fairly protected valley at the edge of desert-like country. In fact, if the residents of nearby ski resorts don't end up in Mexico for the spring off-season, when everything in the mountains is melting snow and mud, they can be found basking in the warmer climates found in southwest Colorado and southern Utah.

Unlike the rest of Colorado, Grand Junction has a 191-day growing season, with mild winters and hot, dry summers. Though its elevation is 4,586 feet above sea level, which may seem high to flatlanders, Grand Junction is surrounded by mountains as high as 12,000 feet to the north and south. To the west stretches scrub, desert and rock, but some

Though the Black Canyon of the Gunnison **opposite** can be run in a rubber raft or kayak, most people prefer to see the deep, dark gorge from the road above.

34

A paleontologist **below left** removes dinosaur fossils from the limestone at Dinosaur National Monument near the Green River. This site has supplied museums all over the world with dinosaur fossil exhibits.

agriculture is possible in those valleys watered by year-round streams.

About 160 million years ago, Grand Junction would have been unrecognizable to present-day inhabitants because it was a tropical valley, occupied by dinosaurs, such as Brontosaurus, Stegosaurus and Brachiosauras. After a few ice ages, bringing us to about 4000 B.C., there were Indians in the area, but we don't know much about them except that they were basketmakers and probably relatives of the Anasazi who occupied the Southwest at about the same time.

In 1776, a group of Spanish explorers led by Friar Escalante and Friar Dominguez went through the area, looking for a northern route to California. As they discovered, this was not easily accessible terrain, and later American settlers bypassed it in favor of the Oregon Trail to the north and the Santa Fe Trail to the south.

The Utes, the last tribe to live in the area, were "removed" after the Meeker Massacre in 1879. The absence of the fierce Utes made

western Colorado a desirable location for white settlers, who moved in quickly and founded Grand Junction, named after the confluence of the Colorado River (then the Grand River) and the Gunnison River, which comes in from the southeast.

In the early twentieth century, Grand Junction was a typical Western town, with a wide and dusty main street, saloons and gunfights, political infighting for power in the new territory and heated rivalries between newspapers competing for limited business. The wide streets are still in evidence in parts of Grand Junction, but the rest of town has given way to the random sprawl typical of a Western city with plenty of room to spread out.

Grand Mesa, which rises up above Grand Junction, is the largest flat-top mountain in the world, spreading over 34,000 acres covered with hundreds of lakes, ponds, meadows, bogs and forests. To ascend a mile to Grand Mesa is to go from the desert into a lush oasis, teeming with wildlife. The mesa began as a lowly plateau and then, about 600 million years ago, gained 300 feet of volcanic lava. Meanwhile, below in the valley, the Colorado and Gunnison Rivers were doing their own carving, leaving fertile river bottom land which has made the orchards of the Grand Valley famous. Today, the Grand Mesa lakes, stocked with Rainbow, Native and Brook trout, attract fly fishermen from all over the world, while the deer, bear and elk attract hunters in the fall. The Powerhorn Ski area and cross-country ski terrain keep Grand Mesa active in the winter. At Land's End, the views stretch into Utah on a clear day.

Colorado National Monument, just west of Grand Junction, offers a look at some spectacular wind-and-water-carved red-rock formations with sheer cliffs, forming bizarre shapes, box canyons and monoliths.

Unaweep Canyon, to the south, may once have been the route of the Colorado River, but now it is a source of many mysteries—it has two mouths, with water running out of both sides, plus a river that actually runs across the canyon. Though a mysterious spot to geologists, both animals and people have found Unaweep Canyon a good place to live for many centuries, as evidenced by the dinosaur fossils and artifacts from Indian civilizations that range from the very ancient to the Utes unearthed there.

High up on Grand Mesas **above**, above the desert floor near Grand Junction, are lush meadows, trout-filled lakes and conifer forests.

Monument Canyon **below** affords spectacular views of the mesas and scrub country that make up western Colorado.

Relic of the frontier gunfights of the Old West, this 'Peacemaker' was modelled in 1873.

The San Juan Mountains of southern Colorado **right** and northern New Mexico contribute many tributaries to the Colorado River.

The Black Canyon of Gunnison

The Gunnison River is the Colorado's largest tributary in the state of Colorado, and it has its own fascinating portrait. It too begins high in the mountains, the San Juans in southern Colorado, near Silverton and Ouray. After contributing to the Blue Mesa Reservoir, the Gunnison charges through the Black Canyon of the Gunnison, a deep gorge with vertical walls of 2,000 feet. The canyon is a perfect example of how powerful the erosion of a stream can be. In this case, the steepness of the river's gradient as it tumbles out of the mountains gives it enough acceleration to chisel faster vertically than horizontally, creating sheer vertical walls and a narrow canyon. The canyon's name is derived from the slate-colored granite through which the water has carved this intimidating chasm. The Black Canyon of the Gunnison can be run on a raft or followed for most of its length by road, with dizzying views from above.

Near the junction of the Gunnison and Umcompahgre Rivers and the town of Montrose was the site of one of the earliest American forts, built to protect soldiers, trappers and explorers from the Ute Indians who then ruled Colorado. The fort was built in the 1830s, and when beaver hats went out of fashion and trappers out of business, it disappeared.

The first formal American expedition to explore the Gunnison was led by a topographical engineer named Captain John Gunnison,

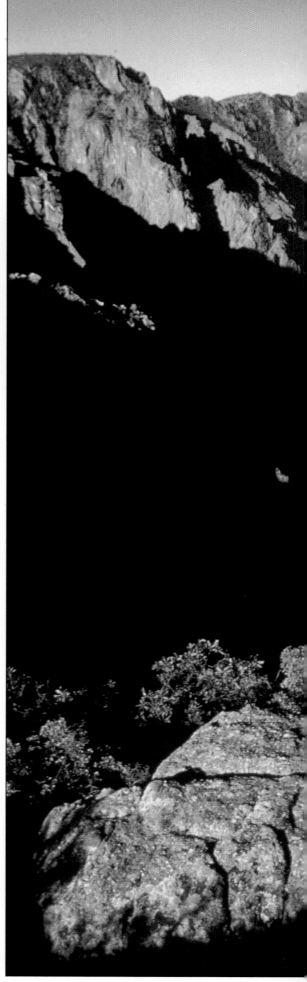

who was mapping possible routes for the Pacific Railway along the 38th Parallel. Captain Gunnison and his party entered southern Colorado via the Santa Fe Trail, crossed the San Juans and descended the western slope of the Continental Divide from Cochetopa Pass (named by the Utes for the buffalo that used it as a trail). As they continued northward, they came to the Gunnison River, which they apparently mistook for the "Grand," and followed it 150 miles to its confluence with the Colorado River, where present-day Grand Junction is. The Gunnison was given the Captain's name after he lost his life on that expedition.

To the west of the Gunnison River is the Uncompahgre Plateau, which forms a headland that the Colorado River winds around on its way to Utah. Just inside the Utah border, the river picks up another large tributary, the Dolores River, named the *Rio de Nuestra Señora de los Dolores* by Father Escalante. The Dolores originates near the Gunnison, but takes the western route through the San Juans and then goes through stark desert and rock country.

Labyrinth Canyon **centre** is an ideal spot from which sightseers can gaze down on the Green River.

Fruit orchards **left** thrive in the mild and sunny climate of the Grand Valley, yet a few thousand feet up are mountains that are covered with snow for much of the year.

39

CANYON LANDS

All the scenic features of this canyon land are on a giant scale, strange and weird. The streams run at depths almost inaccessible, lashing the rocks which beset their channels, rolling in rapids and plunging in falls, and making a wild music which but adds to the gloom of the solitude. The little valleys nestling along the streams are diversified by bordering willows, clumps of box elder, and small groves of cottonwood.
— Major John Wesley Powell

These **globular jars**, made by the Anasazi tribe which dominated the southwest after 1100, have the black-and-white decoration typical of Anasazi ceramic art.

The Northern Canyonlands

The Colorado is a broad, placid river from Moab to its confluence with the Green River. The town of Moab is a jumping-off spot for many who are setting out to explore Southern Utah by river, air or by land, and as such calls itself "the Heart of the Canyonlands.' This is the only town in Utah on the Colorado River, and it has attracted a unique blend of Mormons, employees of the National Park Service, miners, farmers and "river rats," the name given to those who have fallen in love with the river and stuck around it to make a living from it as raft and land tour operators. Though it is surrounded by the majestic La Sal mountains to the east and the tortured desert mazes of the Canyonlands on every other side, Moab has a mild and livable climate, thanks to its low elevation of 4,000 feet.

Moab was settled in 1855 by the Mormons (the Church of Jesus Christ of Latter-Day Saints), who fled persecution in the eastern United States and chose what is now Salt Lake City as their new home—a desert wilderness they could tame and call their own. Thanks to discipline and a unity of purpose, the Mormons quickly turned Salt Lake City into a busy and productive city, and began to spread out and settle other parts of Utah, including Moab. Moab remained quiet and small, barely even on the map, until the United States Atomic Energy Commission (AEC) began its

search for uranium for nuclear weapons. A miner named Charlie Steen was the first to strike uranium south of Moab, in an area the AEC had said was "barren of possibilities." Steen named his mine "Mi Vida" (My Life) and eventually dug out 100 million dollars worth of uranium. In the "uranium rush" that followed Steen's strike, Moab was declared the "Uranium Capital of the World," and became the largest city in Southern Utah.

Canyonlands is not an area one explores casually; planning and preparation must precede even a day hike. Outside of Moab there are no commercial facilities, campgrounds or sources of safe drinking water—in many spots there is no water at all, safe or unsafe. During the summer, temperatures can top 100 degrees, and dehydration can easily strike the unwary. Once you enter the maze of red rocks and canyons off the river, there are few established trails and it is easy to get lost. Good maps are a must. The process of getting in and out, usually by four-wheel drive or by foot, makes Canyonlands one of the most remote spots in the country. Along with the great surge in popularity of exploring Southern Utah and river-running have come thousands of people and a resulting ecological impact. To preserve the river for everyone, the National Park Service has set up guidelines for all those who enter. A permit is recommended for hikers and river-runners, so the Park Service knows where you are and when you should return. Because of unsightly fire rings, fires must be contained within a fire pan. All garbage, trash and human waste must be carried out, and the Park Service obligingly provides directions on how to accomplish this in an efficient and sanitary way.

There are hundreds of prehistoric Indian ruins along this part of the Colorado River. Most are thought to be about 700 years old and were once occupied by the Anasazi and Fremont Indians. Many of the ruins include intact walls and petroglyphs. A petroglyph is a carving or incision on rock.

Arches National Park, just north of Moab, contains the largest concentration of natural arches in the world. Wind and water erosion have carved more than 90 of these formations out of a 300-foot layer of sandstone over the course of 150-million years. A paved road that begins at the park visitor's center provides views of most of the arches, or, even better, a

The prehistoric Anasazi Indians had a highly sophisticated culture, as can be seen in this lovely Punctate brownware bowl.

The sheep and cattle **above** of Mormon settlers have grazed in the Moab valley for over a century.

Below and **Overleaf**
Bizarre rock formations are everywhere in Utah. Arches National Park has the largest concentration of natural arches in the world, carved by wind and water out of a 300-foot layer of sandstone.

series of easy hiking trails give a close-up look.

Across the river from Arches is the Fisher Towers area, which, perhaps better than any other spot along the Colorado, provides a contrast between the mountains and the desert. The Fisher Towers are natural spires. Formations such as Castle Rock, and the Priest and Nuns in Castle Valley give way to cottonwoods and tamarisk along the river and to the green flanks of the mountains to the east. Through breaks in the deep red canyon walls along the river, which rise up above rich bottom lands, there are glimpses of the peaks of the La Sal Mountains.

One of the best viewpoints in the Northern Canyonlands is Dead Horse Point on the west side of the river. This is the end of a promontory called Junction Plateau, which gives views of the river below on both sides, since the Colorado doubles back on itself around what is called a "gooseneck." The gruesome name of Dead Horse Point was bestowed on

Arches National Park is in the heart of red rock country in southeastern Utah. There are more natural stone arches and other strange formations here than in any part of the country. The 300-foot layer of Entrada Sandstone that forms the arches is 150 million years old. The powerful forces of wind, water and frost continue to erode the foundations. Over the centuries, old arches collapse and new ones are formed.

this promontory after local cowboys, using it as a natural corral, let 50 wild mustangs die of thirst, not realizing—or caring—that the animals wouldn't be able to find their way back across the narrow neck.

Dead Horse Point is part of the Island in the Sky District, a triangular plateau between the Colorado and Green Rivers topping sheer cliffs that rise 2,000 feet above the rivers. Other good lookout points on Island in the Sky are Grand View Point and the Green River Overlook.

Crossing the Colorado again, still near the confluence of the Colorado and Green, is the Needles District, yet another area of stone pinnacles rising up to 500 feet from the desert floor in every imaginable size and shape. In Davis, Lavender, Salt and Horse canyons are some of the best spots in Canyonlands for Indian ruins and petroglyphs. This is also a great area for hiking, and Angel Arch is one of the most picturesque sights in Canyonlands.

Just east of Angel Arch is Newspaper Rock Historical Monument, an aptly named sandstone cliff face bearing the inscriptions of visitors and inhabitants spanning 1,000 years. Prehistoric Indians were the first to make their marks on Newspaper Rock, followed by the Utes, and then the white settlers.

Crossing the river to the west once more, the Maze District is located across from the Needles District. This is the most inaccessible region in Canyonlands and one of the most fantastic. On the road to the Maze is Robber's Roost Ranch, once the hideout of Butch Cassidy, the Sundance Kid, and their gang, the Wild Bunch. In Horseshoe Canyon is the "Great Gallery," a large panel of Indian pictographs and petroglyphs. A pictograph is a painting or drawing on rock.

The Confluence

... what a world of grandeur is spread before us! Below is the canyon through which the Colorado runs. We can trace its course for

miles, and at points catch glimpses of the river. From the northwest comes the Green in a narrow winding gorge. From the northeast comes the Grand, through a canyon that seems bottomless from where we stand. Away to the west are lines of cliffs and ledges of rock—not such ledges as the reader may have seen where the quarryman splits his blocks, but ledges from which the gods might quarry mountains that, rolled out on the plain below, would stand a lofty range; and not such cliffs as the reader may have seen where the swallow builds its nest, but cliffs where the soaring eagle is lost to view ere he reaches the summit. Between us and the distant cliffs are ... strangely carved and pinnacled rocks ... Away to the east a group of eruptive mountains are seen—the Sierra La Sal, which we first saw two days ago through the canyon of the Grand. Their slopes are covered with pines, and deep gulches are flanked with great crags, and snow fields are seen near the summits.... Wherever we look there is but a wilderness of rocks— deep gorges where the rivers are lost below cliffs and towers and pinnacles, and then thousand strangely carved forms in every direction, and beyond them mountains blending with the clouds.

—Major John Wesley Powell

Powell's description of the junction of the Green and Colorado Rivers would still be accurate today. This is the spot where two impressive rivers become a giant which has created one of the most awesome canyons on the planet. Just below the confluence, as if to show off its newly gained mightiness, the Colorado has carved out Cataract Canyon, otherwise known as the "Graveyard of the Colorado" for the river-runners' lives it has claimed.

Powell described the beginning of Cataract Canyon this way:

Large rocks have fallen from the walls—great, angular blocks which have rolled down the talus and are strewn along the channel. We are compelled to make three portages in succession, the distance being less than three fourths of a mile, with a fall of 75 feet. Among these rocks, in chutes, whirlpools, and great waves, with rushing breakers and foam, the water finds its way, still tumbling down.

Rock skyscrapers, some as high as 500 feet, dominate the 'Needles District' landscape of the Glen Canyon.

Opposite The words of Major John Wesley Powell are particularly relevant ... "What a world of grandeur is spread before us."

47

The San Juan River, a major
tributary of the Colorado, is known for its
goosenecks, spots where the
river doubles back on itself.

Dead Horse Point, at the end
of Junction Plateau, overlooks the Colorado
River on two sides thanks to
a sharp U-turn the river takes, called a
gooseneck.

Cataract Canyon **right** and **below** below Moab, Utah, and above Lake Powell, is one of the great whitewater challenges left to experienced river-runners on the Colorado.

Near the foot of Cataract Canyon, Powell wrote:

During the afternoon we run a chute of more than half a mile in length, narrow and rapid. This chute has a floor of marble; the rocks dip in the direction in which we are going, and the fall of the stream conforms to the inclination of the beds; so we float on water that is gliding down an inclined plane. At the foot of the chute the river turns sharply to the right and the water rolls up against a rock which from above seems to stand directly athwart its course. As we approach it we pull with all our power to the right, but it seems impossible to avoid being carried headlong against the cliff; we are carried up high on the waves—but not against the rock, for the rebounding water strikes us and we are beaten back and pass on with safety, except that we get a good drenching.

After this the walls suddenly close in, so that the canyon is narrower than we have ever known it. The water fills it from wall to wall, giving us no landing-place at the foot of the cliff; the river is very swift and the canyon very tortuous, so that we can see but a few hundred yards ahead; the walls tower over us, often overhanging so as almost to shut out the light. ... Now that it is past, it seems a very simple thing indeed to run through such a place, but the fear of what might be ahead made a deep impression on us.

At the foot of Cataract Canyon, coming in from the northeast, is the Dirty Devil River, named by Powell and his men: "The water is exceedingly muddy and has an unpleasant odor. One of the men in the boat following ... shouts to Dunn and asks whether it is a trout stream. Dunn replies, much disgusted, that it is 'a dirty devil," and by this name the river is to be known hereafter."

Eight miles downstream from the Dirty Devil is Hite's Landing, named after a prospector who lived there for 15 years. Hite moved away to a more secluded spot in 1898 when the discovery of gold in Glen Canyon attracted a swarm of other prospectors. The small village, once known as a ferry landing, is now a marina on Lake Powell.

The Lost Canyon

The end of Cataract Canyon is also the beginning of Lake Powell, under which lies what was once Glen Canyon. The lake stretches for 186 miles to the Glen Canyon Dam, near the Utah-Arizona border, and has 1,800 miles of shoreline. The 710-foot-high white concrete monolith embedded into the dark red sandstone holds back 27 million acre-feet of water that has flooded a canyon matched only by the Grand Canyon in its magnificence. Looking at pictures of Glen Canyon before it became Lake Powell is enough to evoke the physical pain one feels when a loved one dies, such is the beauty that is gone forever from Glen Canyon.

The building of the Glen Canyon Dam resulted from a passion for massive waterworks projects that grew out of the Reclamation Act, passed by Congress in 1902 to speed up the construction of large-scale irrigation projects to water the dry and thirsty West. In 1922, the seven states that had a claim on Colorado River water—Wyoming, Colorado, Utah and New Mexico, the Upper Basin states, and Arizona, Nevada and California, the Lower Basin states—sat down to hammer out the Colorado River Compact, which would determine which area got how much water. Their

Overleaf The mountains near the Glen Canyon Dam are as forbidding and barren as they look.

BUSINESS REPLY MAIL

FIRST CLASS   PERMIT NO. 604   NEW YORK, N.Y.

POSTAGE WILL BE PAID BY ADDRESSEE

THE

NEW YORKER

25 West 43rd Street
New York, New York 10109

GIVE THE PLEASURE OF READING

SPECIAL GIFT RATES

$32 a year—52 issues—for the first subscription (yours or for someone special). You pay only 62¢ an issue. $20 for each additional one-year gift subscription. Save even more—almost 75% off the newsstand price —and send The New Yorker for only 39¢ an issue.

☐ Start gift subscription only.
☐ Start or extend my subscription only.
☐ Start or extend my subscription plus gift.

Charge my: ☐ American Express ☐ Visa ☐ MasterCard
☐ Payment enclosed ☐ Bill me later.

Account Number _____ Exp. Date _____

Signature _____

Name (please print) _____

Address _____ Apt. _____

City _____ State _____ Zip _____

Gift Recipient's Name _____ (please print)

Address _____ Apt. # _____

City _____

State _____ Zip _____

Gift card to read "From _____ "

Additional Postage: Canada $12.00 a year; other foreign $20.00 a year.
(Please remit in U.S. funds) Basic rate 1 yr. $32.

THE NEW YORKER

4RGN9

first and fatal mistake was overestimating the amount of water available. The amount apportioned to the Lower Basin states was originally about 15 million acre-feet. As it turns out, the average annual amount available is closer to 13.8 million acre-feet. The second mistake the compact made was to conveniently forget Mexico. That country was guaranteed 1.5 million acre-feet annually in 1944, further widening the gap between what the states claimed as their rightful share and the amount actually there.

The Glen Canyon Dam divides the Upper and Lower Basins and provides water and hydroelectric power for central Arizona and a bit of southern Utah and western Colorado. Phoenix, Arizona, uses by far the largest share of Glen Canyon Dam power. The amount of water flowing through the Grand Canyon between Glen Canyon and Hoover dams depends upon how many people in Phoenix are using air conditioners, cooking meals and flushing toilets. As would seem logical, and as anyone running the river regularly will notice, peak usage occurs from 8 A.M. to 10 A.M. and from 4 P.M. to 8 P.M. The more power needed in Phoenix, the higher the river runs, because the more water flowing through the huge turbines in the dam, the more electricity is generated.

Construction of the Glen Canyon Dam began in 1956. The city of Page rose out of the desert to accommodate the thousands of people involved in building this giant, which took seven years to erect. The gleaming white arch, which contrasts sharply with the dark red of the surrounding sandstone, is made of

The Glen Canyon Dam **left** is the first major dam on the Colorado River.

Lake Powell **right** was created when Glen Canyon was filled with water.

Green patches of civilization stand out in the barren red rock country **top** that surrounds the lake.

Though Lake Powell **left** has become a fantastic playground for water sports enthusiasts, as a reservoir it is plagued by problems such as massive silt deposits, evaporation and leaky, absorbent sandstone.

Windmills **right** are common throughout the Southwest, and a welcome sight to thirsty grazers.

4.9 million cubic yards of concrete blocks, and cost 300 million dollars to build, most of which is supposed to be paid back to the federal government by selling electricity.

A number of problems haunt Glen Canyon Dam and Lake Powell, some of them to the glee of conservationists who harbor secret hopes that the dam will crumble, restoring Glen Canyon to the world. Edward Abbey's book, *The Monkey Wrench Gang*, has a large cult following in the Southwest—the theme is a man obsessed with blowing up the dam. Some geologists think the dam may go without the help of dynamite. The sandstone and shale into which the dam is imbedded has a tendency to fall off in blocks and slabs, sometimes enormous blocks and slabs. In spite of 514 rockbolts, some of which go as far as 80 feet into the rock, and in spite of thousands of strain meters, joint meters, resistance thermometers and stress meters, there is no guarantee that the sandstone won't calve like a glacier. Another problem is the porosity of the sandstone, which lets water seep through the

Far beneath the dammed waters
of Lake Powell lies Glen Canyon, once
matched in its grandeur and
beauty only by the Grand Canyon.

dam at rates that can reach 2,500 gallons a minute. What the sandstone doesn't leak out, it absorbs—the shores of Lake Powell soak up an estimated million acre-feet of water a year. Then there is evaporation, which, in the hot desert climate, puts about 450,000 acre-feet of water into the atmosphere every year.

In the spring of 1984, snowpack in the Rockies was 157 percent above average. Normally, when that happens, water is released through spillways to either side of the dam long before Lake Powell reaches its capacity, maintaining a balance that will provide Arizona with electricity without flooding out downstream residents. However, heavy snowfall the year before damaged the spillways, so they weren't available to siphon off excess water, and dam engineers spent that spring holding their breath, hoping the water wouldn't rise any higher—which it didn't—that time.

The harnessing of a force as massive as the Colorado River, no matter how technically perfect, is still subject to the whims of nature—and the bigger the harness, the bigger the potential disasters.

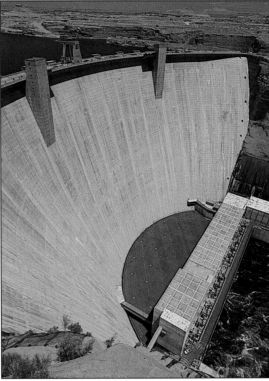

Spillways release water **above** when Lake Powell becomes full.

Visitors to the Glen Canyon Dam **below** can walk along the top of it, and take a tour of the vast labyrinth inside.

Last but not least of Lake Powell's problems are silt deposits that displace 70,000 acre-feet of water a year and threaten to fill up the reservoir in less than two centuries. Whether the weight of all that silt will topple the dam, or whether the dam will become a gigantic waterfall remains a question that will be answered by our descendants.

What really sticks in the craw of conservationists is the government's broken promise to protect Rainbow Bridge National Monument. The drowning of Glen Canyon was a major factor in the first surge of environmental awareness in the early 1960s, and the dam was built with the specific proviso that Lake Powell would not reach Rainbow Bridge. Rainbow Bridge rises 309 feet, tall enough to span the Statue of Liberty, and as the largest natural bridge on earth, it has been declared one of the seven natural wonders of the world. But as the waters rose, it became clear that there would be water under that bridge, threatening its foundations. The alternatives at that point were filling Lake Powell to only half its capacity, or spending millions of dollars to divert the water elsewhere—Congress refused to do either, and today boaters can float underneath the bridge. As the waters of Lake Powell rose to their maximum levels for the first time, other lesser-known but pristine wilderness areas that had been safe on paper were inundated as well.

The "dam fever" that ran rampant throughout the United States in the 1930s, 40s and 50s has abated somewhat, thanks to staunch opposition by conservationists—it was, in fact, only a great hue and cry from the public that prevented the Grand Canyon itself from being turned into yet another "lake."

Rainbow Bridge **left** and **right**, the largest natural bridge on earth, is threatened by the encroaching waters of Lake Powell. Environmentalists believe that the water will eat away at the base of the bridge, causing it to collapse.

The canyons of the Escalante
River are yet another example of incredible
natural beauty drowned beneath
the waters of Lake Powell.

Playground in the Desert

In spite of the tragedy of Glen Canyon, Lake Powell is undeniably a spectacular playground, a paradise for waterskiing, fishing and exploring remote canyons. There are five marinas on the lake where fishing supplies, houseboats, powerboats and sailboats can be rented.

Back on land, on the plateaus above the lake to the east, is an area known as the Trail of the Ancients, a bonanza of historical and anthropological artifacts. The jumping-off point for exploring the area is the small town of Blanding, Utah. In the 500 acres of Hovenweep National Monument are the ruins of six ancient Indian pueblos. Hovenweep is a Ute word meaning "deserted valley," the kind of place many of the peace-loving pueblo cultures gravitated to, seeking a respite from more aggressive and warlike tribes and, later, Spanish conquistadores, missionaries and white settlers.

East of Hovenweep is Natural Bridges National Monument. In a land of superlatives this is yet another—the largest concentration of natural bridges in the world. (Bridges are carved by streams; arches by wind, rain and the sloughing off of loose rock.) The three bridges can all be seen from a paved road or from hiking trails and are called Sipapu,

The favorite haunt of the **red-shouldered hawk** — its red shoulders are often not visible — is broken woodland near lowland rivers.

The incomparable Rainbow Bridge **right**, the world's largest natural bridge.

Red Canyon **above left** is one of hundreds in Glen Canyon Recreation Area, and most have a surprise around every bend in the form of weird rock formations.

An aerial view of Lake Powell **above right** near Hole-in-the-Rock, a narrow road carved out of a 1400-foot cliff by Mormon settlers in 1879.

The Escalante Wilderness Area **below** is a combination of forbidding red-rock desert and lush green valleys. Dozens of tiny, picturesque Mormon farming towns dot the perimeters.

Overleaf Erosion has turned parts of Utah's sandstone into natural cityscapes of towers and minarets.

Capitol Reef National Park **left and right** was named for its rock formations which are white on the top and dome-shaped, reminding some early explorer of the nation's capitol.

Kachina and Owachomo.

To the south, the San Juan river, nearing the end of its journey from the San Juan mountains to the Colorado River, forms a series of convoluted twists and turns, some separated by only a thin wall of stone, named the Goosenecks. The San Juan is a long but little-known river largely because, for a good portion of the year, it is a dry riverbed, with most of its water drawn off for irrigation or backed up behind the Navajo Dam in the Navajo Reservoir in northern New Mexico. The Muley Point overlook is a superb place to get a look at the meanders of the San Juan and the surrounding canyon country. And in the Grand Gulch Primitive Area, where no motorized vehicles are allowed, hikers can explore canyons, Indian ruins and petroglyphs.

On the eastern edge of Paunsaugant Plateau in southwestern Utah, Bryce Canyon National Park **above and right** is famed for its unusual rock formations and brilliant colors. The Pink Cliffs, an escarpment 20 miles long, is perhaps the best-known formation.

The Capitol National Park **left** is part of Waterpocket Fold, a 100-mile long bulge in the earth pocked with basins that catch thousands of gallons of water every time it rains.

Just above the confluence of the San Juan and Colorado rivers is Hole-in-the-Rock, one of the few named features on this part of the Colorado made by man rather than nature. In 1879, a band of Mormon settlers, determined to get to the other side of the river without traveling north or south, spent months camped on the plateau above the river. In that time, they blasted a narrow and treacherous road out of the 1,400-foot cliff. When it was completed, they crossed the river and struggled through the barren wasteland on the other side, finally settling in what became Blanding, Utah.

Horseback trips are a popular
way to visit the colorful amphitheaters and
inner canyons of Bryce Canyon.
On average the cliffs rise to between 8,000
and 9,000 feet above sea level.

THE GREEN RIVER

Like the Colorado, the Green River has its origins in snowmelt that runs off the peaks of the Continental Divide in the Rocky Mountains. High in the Wind River Range in southwestern Wyoming, water trickles down from the Stroud Glacier and surrounding mountains, gathering in a series of lakes as it goes. Technically, the Green is the true source of the Colorado River because it is longer and drains a larger area than the upper Colorado does. But politics played its part, and the Colorado State Legislature simply beat Wyoming to the draw in renaming the Grand River the Colorado. The Indian name for the Green is "Seedskeedee," which is thought to be the name of the prairie chickens found on the plains below the Wind River Range. Father Escalante named it the San Beunaventura, and later Spanish explorers called it the Rio Verde, which was translated to Green River by the Americans.

Today, the Wind River Range has been set aside as the Bridger Wilderness Area, and it is heavily populated by backpackers and climbers who challenge the granite peaks. The beaver are making a comeback, creating pools and bogs along the river. Moose can be seen munching plant growth beside the lakes, and they share this pristone wilderness with other large mammals, such as black bears, bobcats, wolverines, mountain lions and bighorn sheep.

Archaeological evidence found north of Flaming Gorge suggests that Indians of the Yuman culture once lived there, as much as 10,000 years ago, before the last Ice Age. These may have been the ancestors of the present-day Yumans who migrated south when the ice began to invade. Other artifacts dating back to about A.D. 1100 indicate that a Fremont Indian culture also lived there.

The moose, which inhabits coniferous forests, is a protected species, saved from extinction by highly restrictive hunting laws.

Exploring the Green River

The first white man to make a recorded voyage down the Green River was General William H. Ashley, a former military officer who organized a fur company to explore the area and trap beaver. Some of the most famous mountain men in the West joined him, including Jim Bridger and Jedediah Smith.

Ashley's party lost most of their horses to the Crow Indians. To get the pelts out they constructed crude boats made out of buffalo hides stretched over willow frames, called "bullboats," and in August, 1825, prepared to descend the river. This is Ashley's account of these preparations:

I determined to relieve my men and horses of their heavy burdens, to accomplish which, I concluded to make four divisions of my party, send three of them by land in different directions, and with the fourth party, descend the river myself with the principal part of my merchandise.... The partizans were also informed that I would descend the river to some eligible point about one hundred miles below, there deposit my merchandise, and make such marks as would designate it as a place of General Rendezvous for the men in my service in that country, and they were all directed to assemble there on or before the 10th July following.

The spot Ashley picked, a broad and lush meadow valley just above Flaming Gorge, was to be the site of six of these legendary gatherings called Rendezvous, where trappers, traders and thousands of Indians met to indulge in an orgy of drinking, gambling, womanizing, trading and stocking up on supplies before disappearing back into the wilderness for another year. Upon entering Flaming Gorge, Ashley wrote, "We proceeded down the river which is closely confined between two very high mountains ... These mountains present a most gloomy scene. They are entire rock generally of a reddish appearance, they rise the height of from 2 to 4,000 feet.... The rocks that fall in the river from the walls of the mountain make the passage in some places dangerous—windy unpleasant weather."

North of Flaming Gorge is also where two major tributaries, Black's Fork and Henry's Fork, both with their headwaters in the Unita Range of northeastern Utah, come in to meet the Green, swelling its size to an impressive river. The terrain surrounding the Green changes abruptly at Flaming Gorge from flatlands to mountains as it cuts through the

Opposite The Colorado River irrigates hundreds of small farms as it meanders **74** through Western Colorado.

Unitas. Though the entrance to Flaming Gorge is forbidding enough to have discouraged many river travelers and caused others to wonder whether they were going to be sucked into a hole in the mountains, the real rapids don't begin until Red Canyon. Flaming Gorge was later described by one of Powell's men as, "A pass through the rocks which are of red sandstone and look very much like a flame in the rays of the sun." Thus it was named. The spectacular rocks are much less in evidence today, hidden underneath the waters of the Flaming Gorge Reservoir.

After struggling through the rapids of Red Canyon, Ashley could have left the river at Brown's Hole, above Lodore Canyon, as he originally planned. But he seems to have been irresistibly drawn to explore the mysterious canyons below. They struggled south, hauling the boats over rocks much of the way, and by the time they emerged, just north of Desolation Canyon, they were half-starved and ready to leave the river. Ashley bought some horses from the Indians, returned to the Rendezvous, and in the fall arrived in St. Louis with 9,000 pounds of beaver pelts.

Nearly 50 years later, Major Powell began his famous river trip near Green River, Wyoming. He was also enchanted with these canyons: "Each step is built of red sandstone, with a face of naked, red rock, and a glacis clothed with verdure. So the amphitheatre seems banded red and green, and the evening sun is playing with roseate flashes on the rocks, with shimmering green on the cedars' spray, and iridescent gleams on the dancing waves. The landscape revels in sunshine."

The gloomy entrance to Lodore Canyon, called the Gate of Lodore, was a good indication of what lay ahead for Powell's party:

During the afternoon we come to a place where it is necessary to make a portage. The little boat is landed, and the others are signaled to come up. I walk along the bank to examine the ground, leaving one of my men with a flag to guide the other boats to the landing place. I soon see one of the boats make shore all right and feel no more concern; but a minute after I hear a shout, and looking round, see one of the boats shooting down the center of the sag.

It is the No Name, *with Captain Howland, his brother, and Goodman. I feel that its going*

over is inevitable ... The first fall is not great, only ten or twelve feet, and we often run such; but below, the river tumbles down again for forty or fifty feet, in a channel filled with dangerous rocks that break the waves into whirlpools and beat them into foam. I pass around a great crag just in time to see the boat strike a rock, and, rebounding from the shock, career

As the Green River **left** approaches its confluence with the Colorado, it plunges through barren canyons filled with rocky rapids. Boat trips led by experienced river-runners are a safe and exciting way to experience the rapids.

and fill the open compartment with water. Two of the men lose their oars; she swings around, and is carried down at a rapid rate, broadside on, for a few yards, and strikes amidships on another rock with great force, is broken quite in two, and the men are thrown into the river; the larger part of the boat floating buoyantly, they soon seize it, and down the river they drift, past the rocks for a few hundred yards to a second rapid, filled with huge boulders, where the boat strikes again, and is dashed to pieces, and the man and fragments are soon carried beyond my sight. Running along, I turn a bend, and see a man's head above the above, washed about in a whirlpool below a great rock.

Great upheavals in the earth's
crust at Dinosaur National Monument have
exposed layers of geological
time and a wealth of dinosaur fossils.

The Green River has many moods.
It can be a placid stream winding through
fields or between high, dark
cliffs, or it can abruptly plunge down to rapids
guaranteed to produce a surge
of adrenalin in the most intrepid rafter.

Opposite The word Colorado is Spanish for "red river," a name that still applies above the Glen Canyon Dam, where it carries huge amounts of silt from the surrounding red shale and sandstone.

The Green River **above** gets its clear, icy water from glaciers in the Wind River Range of Wyoming.

The Green River **left** twists through Labyrinth Canyon in Utah before it joins with the Colorado River in Canyonlands National Park above Cataract Canyon.

It is Frank Goodman, clinging to it with a grip upon which life depends. Coming opposite, I see Howland trying to go to his aid from an island on which he has been washed. Soon, he comes near enough to reach Frank with a pole, which he extends toward him. The latter lets go the rock, grasps the pole, and is pulled ashore. Seneca Howland is washed farther down the island, and is caught by some rocks, and, though somewhat bruised, manages to get ashore in safety.

It took them nine days to get through the 21 miles of Lodore Canyon. At a rapid they named Hell's Half Mile another boat was almost lost, and cinders from their campfire started a fire that destroyed much of their bedding and cooking utensils. When they finally reached a quiet stretch of water where the Yampa River joins the Green, Powell wrote: "This has been a chapter of disasters and toils, notwithstanding which the Canyon of Lodore was not devoid of scenic interest, even beyond the powers of the pen to tell. The roar of its waters was heard unceasingly from the hour we entered it until we landed here. No quiet in all that time. But its walls and cliffs, its peaks and crags, its amphitheatres and alcoves, tell a story of beauty and grandeur that I hear yet—and shall hear."

What Powell didn't know was that he was sitting on one of the largest deposits of dinosaur fossils in the world. Earl Douglass, working for the Carnegie Museum, stumbled on this find in 1909, and it was later made a national monument. The remains of these huge creatures, which lived about 130 million years ago, have been shipped to museums all over the world, and scientists are still excavating. The visitor center at Dinosaur National Monument is built over the quarry so that visitors can look over the shoulders of the paleontologists who work there.

The sections of the Green and Yampa rivers within Dinosaur National Monument still provide some of the best river running in the west. The area is surrounded by badlands, a tortured maze of inhospitable desert and scrub once notorious as the hideout of outlaws.

From here the Green River, having originated in Wyoming and passed through a corner of Colorado, begins to head south through Utah. It flows through the Ouray National Wildlife Refuge, and a section of the Uintah and Ouray Indian Reservation. After picking up the Uinta, Duchesne and White Rivers, it takes another plunge into Desolation Canyon. This is more barren country, without even the lush riverside growth and oases which grace the river to the north. It picks up two rivers coming in from the west, the Price and the San Rafael, before heading into Labyrinth Canyon and the confluence with the Colorado River.

THE NAVAJO PEOPLE

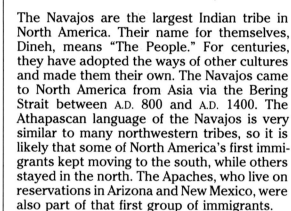

*In beauty I walk.
With the pollen of dawn upon my path I
wander.
With beauty before me, I walk.
With beauty behind me, I walk.
On the trail of morning, I walk.*

—Navajo Song

The Navajos are the largest Indian tribe in North America. Their name for themselves, Dineh, means "The People." For centuries, they have adopted the ways of other cultures and made them their own. The Navajos came to North America from Asia via the Bering Strait between A.D. 800 and A.D. 1400. The Athapascan language of the Navajos is very similar to many northwestern tribes, so it is likely that some of North America's first immigrants kept moving to the south, while others stayed in the north. The Apaches, who live on reservations in Arizona and New Mexico, were also part of that first group of immigrants.

Early Navajos were a primitive, nomadic culture, but they soon began learning the ways of the Pueblo Indians, whose peaceful villages they raided. The Pueblos taught the Navajos how to plant corn, squash and other crops, and how to weave. When the Spaniards introduced horses and sheep to North America, the Navajo were among the first to take advantage of them. Both have become a valuable part of Navajo culture—the sight of a Navajo woman herding sheep below statuesque red rocks has become a stereotype. Navajo rugs woven from the wool of these sheep fetch high prices at the run auctions held on the reservation.

After long decades of Indian wars, the whites began to get the edge through sheer force of numbers and superior weapons. In 1862, Kit Carson, on behalf of the United States Army, offered the Navajo the chance to surrender. They chose to fight, and Carson drove them into Canyon de Chelly, a 1,000-foot-deep gorge carved out of red rock. The valley floor had long been occupied by Navajos, and they hid in caves in the cliffs.

Eventually Carson realized he couldn't beat them in their stronghold, so he burned their fields and orchards. It took three years, but the Navajo were finally forced to surrender or starve to death. Once Carson had rounded up the 7,000 men, women and children, they were forcibly marched to Bosque Redondo, Okla-

homa, a parched wasteland 300 miles away. Many died along the way, on what is known as "The Long Walk," and once they got there, survival was virtually impossible. Five years later, the United States government began to feel pangs of guilt over what would be genocide if it continued, and the Navajos were offered rich farm country in Arkansas. They weren't interested; they wanted to go back home. The Navajo leader, Barboncito, and other clan leaders convinced General William T. Sherman to let them return to their land, saying, "Our grandfathers had no thought of living in any other country than our own and it is not right we should abandon it. Here we plant, but the soil does not yield. All our animals have died. We have nothing left in the way of possessions but a gunnysack to wear during the day and to cover us at night. It makes my mouth dry and my head hangs low seeing us die here. If we are taken back to our own country we will call you father and mother. If there was only a single goat there, we would all live off of it."

General Sherman assured President Andrew Johnson that the land the Navajos wished to return to was "as far from our future possible wants as was possible to discover," and the government drew up a treaty that allowed the Navajos three million acres in northeastern Arizona and a chunk of northwestern New Mexico.

As it turned out, the Navajo reservation contained vast deposits of surface coal, uranium, natural gas and timber. The rights to exploit these natural resources have been leased from the Navajos by large corporations, and most of the profits go to them, not the Navajos. Strip-mining has ruined thousands of acres of grazing land and claims much of the already scarce water; uranium tailings have leaked into groundwater supplies and are suspected of creating a high cancer rate; and smoke from coal-burning power plants pollutes the air. Meanwhile, the rate of Navajo unemployment is among the highest in the country, life expectancy is ten years less than the national average and tuberculosis is still a major health problem. Navajo children have to leave the reservation to go to boarding schools run by the Bureau of Indian Affairs. The conflicts between Navajo culture and white culture, which are almost diametrically opposed, have created rampant psychological

The **turkey**, hunted for food by Indians long before the arrival of the Europeans and still abundant in southwestern woodlands, gets its name from its 'turk-turk' call.

This Navajo hogan **opposite** may be deserted because the family has taken their sheep to better grazing land for the summer. Some traditional Navajos still abandon their hogan permanently when someone dies inside. All hogan doors face the rising sun.

problems among the Navajo and contributed to a high rate of alcoholism.

The Navajos traditionally lived in wood and mud structures called hogans, usually round or octagonal, always with the door facing east. Though many now live in housing developments and mobile homes, there are still hogans scattered about. Many older Navajos speak very little English and have a minimum of contact with white culture, though occasionally a TV antenna can be seen next to a hogan's smokehole and there is nearly always a pickup truck parked in front.

Land of The Navajo

I see the Earth
I am looking at Her and smile
Because She makes me happy.
The Earth, looking back at me
is smiling too.
May I walk happily
And lightly
On Her.

—Navajo Song

Like much of the Southwest, Navajo Land is red-rock country, and some of the most impressive formations are found on the reservation. To drive through the reservation and see the landscape through the eyes of the Navajo is to understand their way of life better.

"I hope to God you will not ask us to go to any country but our own," said Barboncito. "When the Navajos were first created, four mountains and four rivers were pointed out to us, inside of which we should live, and that was to be Dinetah. Changing Woman gave us this land. Our God created it especially for us." This is the Holy Land for the Navajo, in much the same way that Jerusalem is for Christians, Jews and Muslims, with every landmark filled with religious meaning.

The four sacred mountains correspond to the four directions. Mount Taylor looms up on the horizon to the south, snow-capped much

of the year. In the Navajo language it is called Tso Dzil, or Tourquoise Mountain. Tourquoise Girl lives on the mountain watching over Dinetah—the People's Country—and is in turn guarded by Tliish Tsoh, the Big Snake. The black-lava flows at the foot of the mountain are the blood of one of the evil monsters slain by the Hero Twins, who helped make Dinetah safe for the Navajo to live in. To the north is Mount Hesperus; to the east Mount Blanca;

Above For the Navajo Indians, sheep are a form of currency, a source of food, warmth and the wool that is woven into their beautiful blankets.

Below This is a classic combination in the Southwest: strangely carved red rocks, and an Indian on horseback tending the family herd of sheep.

Every major rock formation
in Monument Valley Tribal Park **above** has
religious meaning for the Navajo
People. Left Hand and Right Hand Mittens **left**
are especially beautiful at sunrise.

Navajo silver jewelry **above**, such as this traditional Squash Blossom necklace, is highly valued for its elegant simplicity of design.

Navajo women originally wove their colorful rugs as blankets **centrefold** for warmth; today they can fetch thousands of dollars at auctions, trading posts and galleries all over the world.

and to the west is San Francisco mountain. Each has its own legends.

Shiprock stands alone in the scrub country, looming 1,450 feet into the sky. It is actually the core of a volcano, worn down to jagged pinnacles by centuries of erosion. There are a number of Navajo myths surrounding this massive landmark. In one, its cliffs were the aerie of a winged monster that was terrorizing the Dinetah. One of the Hero Twins killed the monster, and was lowered off the rock in a basket woven by Spider Woman.

Spider Woman's rock is a tall, thin red pinnacle rising by itself out of the valley floor in Canyon de Chelly. It was she who taught the Navajos how to weave their beautiful rugs.

There are many other sights to see in Canyon de Chelly and Monument Valley, both within the reservation. Navajos still live in Canyon de Chelly, growing fruit and vegetables on the valley floor and living a traditional life.

The Hopi: People of Peace

The white man, through his insensitivity to the way of Nature, has desecrated the face of Mother Earth. The white man's advanced technological capacity has occurred as a result of his lack of regard for the spiritual path and for the way of all living things. The white man's desire for material possessions and power has blinded him to the pain he has caused Mother Earth by his quest for what he calls natural resources. And the path of the Great Spirit has become difficult to see by almost all men, even by many Indians, who have chosen instead to follow the path of the white man. We have accepted the responsibility designated by our prophecy to tell you that almost all life will stop unless men come to know that everyone must live in peace and harmony with Nature and with each other. Only those people who know the secrets of Nature, the mother of us all, can overcome the possible destruction of land and life.

—Message from the Hopi Elders to the United Nations

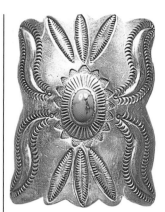

Above This wrist guard is a fine example of the Navajo tribe's traditional craftsmanship in silver jewelry.

Kachina dolls are not playthings,
but highly religious symbols of Hopi gods who
emerge from the San Francisco
Mountains to teach their people the Hopi
Way.

Betatkin Ruin **left** at Navajo National Monument is one of hundreds of Pueblo cities in the Southwest mysteriously abandoned by their inhabitants centuries ago.

Ancient Hopi symbols and designs make their exquisite pottery **right** come alive.

The Hopi reservation is a square in the middle of the Navajo reservation, much smaller, but as much the ancestral home of the Hopis as Canyon de Chelly is for the Navajos. The Hopis are Pueblo Indians, a very different people from the Navajos. Their name for themselves means 'the Peaceful People." The Hopi and Navajo are traditional enemies, and today they live close to each other rather grudgingly. While the Navajos have the largest population of any tribe in America, the Hopis have the most intact culture. They tend to be religious, conservative and do not readily accept outsiders. Their culture is inseparable from their religion, which is based on a deep reverence for the earth and staying in tune with it. The Hopi threw more than one overzealous Spanish missionary over the cliffs of their mesas, and resisted white culture and religions long after other tribes.

Most Hopis live on First, Second or Third Mesa, where they have been since at least A.D. 1100—they say they have been there for 30,000 years, and recent archaeological evidence suggests that this is true. Oraibi, a tiny village perched on a small mesa with steep sides above Third Mesa, is the oldest continuously inhabited settlement in North America. The Hopis have been farming the dry land around their mesas for centuries, and the sacred corn is their most important crop. They believe the reason they were given the arid land to farm is so they will stay in touch with Mother Earth through their constant need for rain. Most of their ceremonies revolve around rain. As in most pueblos in the Southwest, the center of their ceremonials is the kiva, a sacred underground chamber.

The Hopi family is matriarchal, with property passed on through the mother, and relationships are defined by a clan system based upon matrilineal descent. Land is apportioned by the clan, though technically it is controlled by the whole tribe. Hopi women are supposed

The Petrified Forest in Arizona **above** is evidence that there was once a great forest in what is now a desert.

Hopi pottery **below** is highly coveted among collectors of Indian art.

to build their homes, because in the event of a divorce they keep the house and furniture.

The most unique Hopi craft is the carving and decorating of Kachina dolls, which are used in ceremonies and to teach children Hopi legends and ways. These figurines can be any-where from a few inches to several feet high, and may be simply painted or adorned with feathers, beads and clothing. The Hopi create exquisite baskets and pottery. Hopi jewelry is also unique, usually bracelets, necklaces and earrings of silver overlay, with a sophisticated simplicity of design. Both the Hopi and the Navajo have many fine painters in their midst who have gained national recognition, and whose work is coveted by collectors.

Overleaf Geologic history comes alive in the exposed strata of Grand Canyon National Park, but many mysteries still remain.

PART II
THE
GRAND CANYON

INTRODUCTION

In the Grand Canyon, Arizona has a natural wonder which, so far as I know, is in kind absolutely unparalleled throughout the rest of the world. I want to ask you to do one thing in connection with it in your own interest and in the interest of the country—to keep this great wonder of nature as it now is.... I hope you will not have a building of any kind, not a summer cottage, a hotel or anything else, to mar the wonderful grandeur, the sublimity, the great loveliness and beauty of the canyon. Leave it as it is. You cannot improve upon it. The ages have been at work on it, and man can only mar it. What you can do is to keep it for your children, your children's children, and for all who come after you, as one of the great rights which every American if he can travel at all should see. We have gotten past the stage, my fellow citizens, when we are to be pardoned if we treat any part of our country as something to be skinned for two or three years for the use of the present generation, whether it be the forest, the water, the scenery. Whatever it is, handle it so that your children's children will get the benefit of it.

—President Theodore Roosevelt
May 6, 1903 at the Grand Canyon

President Roosevelt did not completely get his request fulfilled—there are commercial facilities at the North and South Rims of the Grand Canyon, and the Park Service has chiseled trails and viewpoints out of the cliffs. However, Roosevelt did declare the Canyon a National Monument in 1908, and in 1919 Congress created Grand Canyon National Park. The park encompasses 1,900 square miles and about 280 miles of the Colorado River.

For an estimated six million years, the forces of water, wind, heat, cold and upheavals from beneath the earth have carved this mile-deep chasm out of a mountain called the Kaibab Plateau. The cliffs of the Grand Canyon provide a layer-cake look at the evolution of the earth's crust—starting with the Kaibab limestone at the rim, a mere 200 million years old, down to metamorphic rock, ancient granite and schist at the bottom, which may be 2 billion years old.

Before the Glen Canyon Dam was built, the Colorado carried anywhere from one million to 55 million tons of the red silt that gave it its name through the Grand Canyon every day, a virtually incomprehensible amount of earth moving. Now most the silt is left behind the dam, filling up Lake Powell instead, and the water runs green on most days.

Not too far up from the river bottom, a layer of sandstone, shale and limestone called the Unkar-Chuar group contains fossils which are evidence of the first life on the planet. Skipping up millions of years to the Tonto group of shale and sandstone is the first evidence of animals with backbones, and then, getting near the top in the Supai formation, are the first land-dwelling animals.

Archaeological evidence indicates that Indians have lived in the American Southwest for almost 30,000 years. The earliest artifacts found in the Grand Canyon date back about 4,000 years, and are animal effigies made out of split willow twigs. These animal forms— deer, bison, bighorn sheep—were found in Lukas Cave near Phantom Ranch. The people who placed the figurines in the cave and carefully covered them with a cairn of rocks remain a mystery, and there is no further evidence of human presence in the Canyon until about 500 B.C., when the Anasazi arrived. Also known as the Basketmakers, the Anasazi hunted with snares, spears, spearthrowers and later, bow and arrow. They cultivated corn, beans and squash and collected cactus fruits, agave, mesquite beans, piñon nuts and other wild food.

In addition to their beautiful baskets, which were usually woven out of yucca, the Anasazi made pottery, which was used for cooking and storage. By A.D. 300, these Indians had evolved into what anthropologists call the Pueblo Group. The ruins of their dwellings are scattered throughout the Southwest. There are more than 2,000 known pueblo ruins in the Grand Canyon, most of which were probably occupied between A.D. 1050 and A.D. 1150. Then something happened throughout the Southwest that caused most of the major pueblos to be abandoned—probably a prolonged drought, but there is little agreement among archaeologists about what exactly happened. Around 1300, the ancestors of the present-day Havasupi and Hualapai moved into the Grand Canyon, with a simpler culture than the Anasazi who had preceded them.

Opposite Though the water level of the Colorado River sometimes gets low during a dry summer, it's usually dictated on a day-to-day basis by how much electricity residents of Southwestern cities are using.

96

THE GREAT CHASM

The Hopi Indians made annual visits to Sipapu, a mineral spring in the Grand Canyon that they believe is the spot where life emerged from the underworld and where the dead return. Every year, they paid their respects, gathered ceremonial salt and returned to their mesas in the east. Captain Garcia Lopez de Cardenas, leader of one of the parties sent by the Spanish conquistadore Coronado to find the legendary Seven Cities of Cibola, met the Hopis in 1540. They told him there was a great river to the west, but that was all. Cardenas and his men were the first whites to peer over the rim of the Grand Canyon. One of Coronado's men later wrote, "It seemed to be more than three or four leagues in an air line across to the other bank of the stream that flowed between them. The country was elevated and full of low, twisted pines, very cold, and lying open to the north. They spent three days on this bank, looking for a passage down to the river, which looked from above as if the water was six feet across, although the Indians said it was half a league wide. It was impossible to descend." Some of Cardenas' men spent a frustrating three days trying to find their way to the bottom, without success.

The conquistadores were followed by the padres, and more than 200 years later in 1776, Father Francisco Silvestre Velez de Escalante set out from Santa Fe to find a northerly route to the Spanish missions in California. He and his men spent nearly four months wandering among the impassable canyons and gorges of the Colorado and Green Rivers. They finally managed to find what is now the Lee's Ferry crossing on the Colorado River, but by that time they had given up trying to find a northerly route and were just trying to find their way home. Before they reached Lee's Ferry, Escalante wrote: "The River Colorado flows along here from north-north-east to south-south-east, very deep, with high banks, so that if one should cultivate the land on the banks of the river, although the soil might be good, the stream would be of no service to him. We caught sight this afternoon of the precipices lining the sides of the river, and seen from the western side they resembled a long ridge of houses."

Of their journey from the Paria plateau to the mouth of the Paria River and Lee's Ferry on the Colorado (which he also called the Cosninas) Escalante wrote:

We pushed ahead for three leagues and a half, and arrived at the spot that we thought might be the outlet of the valley. It is a corner entirely surrounded by mountains and peaks, very lofty, of colored red earth of different formations, and as the soil underneath the surface is of the same color, it has an agreeable aspect. We continued to the same direction, traveling with great difficulty, for the horses sank to their knees in the soft earth, when the surface was broken through. Having covered another league and a half, we reached the great river of the Cosninas. Another smaller one unites with it at this point, and we called this the Santa Teresa. We crossed this one, and pitched our camp on the larger one close to a precipice of gray stone.

The spot where they were finally able to cross the Colorado, after many attempts, was just above Lee's Ferry, known as the Crossing of the Fathers.

The 40-mile, five-day hike through the narrow, 1,200-foot cliffs of Paria Canyon has become a favorite of Canyonlands regulars. Since there is no way to the top from beginning to end, not many make the whole distance, but even a day hike up a few miles and back is worthwhile.

After the Escalante expedition, the Grand Canyon was largely avoided by explorers and trappers, who saw it only as an obstacle. James Ohio Pattie probably crossed the Colorado at Lee's Ferry in 1827, in the process of following the Colorado from the Gulf of California to its headwaters, but he was not a topographer, and his accounts of his journey tend to focus more on encounters with Indians than on descriptions of the landscape.

In 1857, Lieutenant Joseph C. Ives attempted to navigate up the Colorado River in a steamboat. He was forced to send the boat back south at the Black Canyon, but he continued overland to the Grand Canyon. His reaction to it was: "The region last explored is, of course, altogether valueless. It can be approached only from the south, and after entering it there is nothing to do but leave. Ours has been the first, and will doubless be the last, party of whites to visit this profitless locality. It seems intended by nature that the Colorado River, along the greater portion of its lone and majestic way, shall be forever

Lee's Ferry, Arizona, **opposite** is one of the few spots on the Colorado River where early explorers were able to cross.

98

Archaeologists have unearthed treasures such as this duck-shaped **pottery jar** from the pueblo ruins found all over the Southwest.

unvisited and and undisturbed."

Ives would have been amazed to know that a hundred years later millions of people a year from all over the world are going out of their way to see the Grand Canyon, and that about 15,000 people a year descend the river in boats.

It is possible that a prospector named James White inadvertently floated down the Grand Canyon on a raft in an attempt to escape the Indians who had killed his companions, but most historians feel that such a feat would have been impossible, and that White didn't know where he was. Be that as it may, the journey of Major John Wesley Powell was the first recorded descent, and his journal remains a classic piece of Grand Canyon literature.

Powell enlisted in the Union army at the age of 20, and he lost his right arm at Shiloh. He later became a professor of geology at Wesleyan University, Illinois. He became interested in exploring the river during a vacation in Colorado and Wyoming. Though Powell later organized the United States Geological Survey, it was academic institutions that funded his expedition down the Grand Canyon. He had four 21-foot rowboats built, and recruited nine men to accompany him. With provisions to last ten months, and scientific instruments for recording what they saw, the men set out down the Green River from the town of Green River, Wyoming on May 24, 1869. By the time they reached the Grand Canyon on August 13th, they had lost two of four boats, and one man had deserted the expedition.

Twelve miles below the Glen Canyon Dam is Lee's Ferry, the famous crossing named after the renegade Mormon, John D. Lee, who operated a ferry there while hiding from the law. The Mormons who settled in Salt Lake City had become increasingly angered by a United States government that would not recognize Utah as a state because of the Mormon practice of polygamy and was threatening to send in the Army to control the Mormons. The Mormon settlements in Southern Utah complained bitterly to their superiors in Salt Lake City of abuse and persecution by settlers passing through in wagon trains. When the Francher party, a wagon train of about 140 men, women and children, came through, the atmosphere was tense. The wagon train was

Previous page The mighty Colorado River becomes a tiny blue ribbon when seen from the high walls of the Grand Canyon.

Tusayan Ruins on the South Rim of the Grand Canyon **above** takes the visitor back 800 years, to the homes of prehistoric canyon dwellers.

Millions of people from all over the world visit the Grand Canyon **left** each year.

Baloney boats **right** are a clumsy but safe form of transport through the rapids of the Grand Canyon.

forced to camp in Mountain Meadows, at one end of a Mormon ranch, to fight off attacking Indians. Instead of helping to fight off the Indians, Major John D. Lee, an adopted son of Mormon leader Brigham Young, cooperated with them. Under the guise of helping the wagon train, he and his mob drew them out of their protected circle and opened fire on them at close range. They then looted the wagon train and fled, leaving the Indians to finish the job of massacring 120 men, women and children.

Meanwhile, Brigham Young had sent a message south ordering his men to let the wagon train pass through peacefully, but it got there too late, and the men who participated in the massacre agreed to blame it on the Indians. The truth was too awful to be hidden, and Lee fled to the most remote spot he could think of, the Grand Canyon. Twenty years later, he was executed for his crime, but not before he had lent his name to the crossing. Today, Lee's Ferry is known as the starting point for most river trips down the Grand Canyon. Just four miles below Lee's Ferry is the Navajo Bridge, a steel arch spanning the canyon 467 feet above.

Trains of **covered wagons** carried settlers and gold-prospectors across hostile Indian territory to the safer haven of the far West.

INTO THE DEPTHS OF THE EARTH

*P*owell learned a lot of geology during the arduous months traveling down the Green River. He had undertaken the journey to make a scientific mapping expedition, and to find out if the tales of impassable waterfalls and the river disappearing into the bowels of the earth were true.

These canyon gorges, obstructing cliffs, and desert wastes have prevented the traveler from penetrating the country, so that until the Colorado River Exploring Expedition was organized it was almost unknown. In the early history of the country Spanish adventurers penetrated the region and told marvelous stories of its wonders. It was also traversed by priests who sought to convert the Indian tribes to Christianity. . . . Yet enough had been seen in the earlier days to foment rumor, and many wonderful stories were told in the hunter's cabin and the prospector's camp—stories of parties entering the gorge in boats and being carried down with fearful velocity into whirlpools where all were overwhelmed in the abyss of waters, and stories of underground passages for the great river into which boats had passed never to be seen again. It was currently believed that the river was lost under the rocks for several hundred miles. There were other accounts of great falls whose roaring music could be heard on the distant mountain summits; and there were stories current of parties wandering on the brink of the canyon and vainly endeavoring to reach the waters below, and perishing with thirst at last in sight of the river which was roaring its mockery in their ears.

As they approached what Powell named Marble Gorge, he knew that the most difficult part of the journey was beginning. This was the entrance to the Grand Canyon. He wrote: "With some feeling of anxiety we enter a new canyon this morning. We have learned to observe closely the texture of the rock. In softer strata we have a quiet river, in harder we find rapids and falls. Below us are the limestone and hard sandstones which we found in Cataract Canyon. This bodes toil and danger." Powell's premonition was all too correct, and in the next 287 miles he and his men dropped 2,000 feet, over more than 150 rapids.

What Powell euphemistically called marble is actually Redwall limestone, which forms long stripes of multicolored rock along much of the Canyon. Though he was not lulled into a false sense of security, Powell was enraptured with the scenery in Marble Gorge: "The river turns sharply to the east and seems enclosed by a wall set with a million brilliant gems. On coming nearer we find fountains bursting from the high rock overhead, and the spray in the sunshine forms the gems which bedeck the wall. The rocks below the fountain are covered with mosses and ferns and many beautiful flowering plants. We name it Vasey's Paradise, in honor of the botanist who traveled with us last year."

Other sights in Marble Gorge include Redwall Cavern, a natural amphitheater which Powell estimated could hold 50,000 people (he was exaggerating), and Buck Farm Canyon, which is so narrow in some spots that one can stand in the middle and touch both sides.

Though the Grand Canyon geographically begins at the head of Marble Canyon, the spot where the Little Colorado joins the Colorado from the east, restoring some of its natural reddish color, is where the Grand Canyon truly begins—there is no doubting that the river is about to enter a great chasm. In the most often quoted passage of Powell's journey, he writes:

We are now ready to start on our way down the Great Unknown. Our boats, tied to a common stake, chafe each other as they are tossed by the fretful river. They ride high and buoyant, for their loads are lighter than we could desire. We have but a month's rations remaining. The flour has been resifted through the moquito-net sieve; the spoiled bacon has been dried and the worst of it boiled; the few pounds of dried apples have been spread in the sun and reshrunken to their normal bulk. The sugar has all melted and one on its way down the river. But we have a large sack of coffee.

We are three quarters of a mile in the depths of the earth, and the great river shrinks into insignificance as it dashes its angry waves against the walls and cliffs that rise to the world above; the waves are but puny ripples, and we but pigmies, running up and down the sands or lost among the boulders.

We have an unknown distance yet to run, an unknown river to explore. What falls there are,

Marble Canyon, whose 2,500-feet high walls are a pageant of multi-colored limestone, marks the entrance to the Grand Canyon.

Opposite The narrow band of the Colorado River in the depths of the awesome canyon it has carved for itself is a vivid illustration of the power of erosion.

Above Phantom Ranch, seen from the South Rim across Tonto Plateau, is the only sign of civilization for the length of the Colorado in Grand Canyon.

In spring the ubiquitous tamarisk tree **right** blooms all along the lower Colorado River. Originally imported from the Far East to stop erosion, it has now become a pest.

we know not; what rocks beset the channel, we know not; what walls rise over the river, we know not. Ah, well! we may conjecture many things. The men talk cheerfully as ever; jests are bandied about freely this morning; but to me the cheer is somber and the jests are ghastly.

As the river makes a great loop south through Granite Gorge, there are many rapids—Unkar, Nevills, Hance, Sockdolager, and Grapevine. Side canyons include Asbestos, Vishnu, Lonetree, Clear Creek, Zoroaster and Cremation—some with creeks that draw their water from underground springs or some that come to life only when rain funnels down from above. They remain largely unexplored by the thousands of people who float past them every year.

River-runners tend to talk about the Grand Canyon in terms of rapids and miles—Lee's Ferry is mile zero, the entrance to Lake Mead at the other end is mile 277. The Phantom Ranch, at mile 87, is the only sign of civilization—telephone lines, a ranger station and a hostel—in the Canyon. From the South Rim, the Kaibab Trail winds down the cliffs to the river, well populated for much of the year with hikers and mules carrying passengers and gear. The Kaibab Suspension Bridge that spans the Canyon here is the only crossing on the river between Navajo Bridge and the Hoover Dam, and it is only a footbridge. This is also where Bright Angel Creek joins the Colorado River, and the Kaibab Trail continues along it up to the North Rim of the Canyon.

Powell named the Bright Angel, a creek with clear water, probably in response to having christened upstream tributaries with such names as the Dirty Devil and Desolation Canyon. He is responsible for many of the Canyon's names. Clarence Dutton was a member of the U.S. Geological Survey who explored the Grand Canyon a few years after Powell did and made the first accurate geological maps. He bestowed many of the buttes, spires, pinnacles and other strange rock formations in the Canyon with names drawn from Eastern religions: Hindu Amphitheater, Vishnu Temple, Krishna Shrine, Shiva Temple, to name a few. This religiosity must have seemed appropriate to later geologists and topographers who were faced with giving names to the awesome formations, and they drew from

other mythologies and religions: Wotan's Throne, Venus Temple, Solomon Temple, Siegfried Pyre, Tower of Set, Cheops Pyramid and so forth. Still others bear names the Indians gave them, such as Awatubi Crest, Nankoweap Mesa and Cheyava Falls.

Many of the rapids in the Grand Canyon were formed by the flash floods which can bring massive chunks of rocks off the cliffs, down from the side canyons, or simply push them down the riverbed and drop them, creating the obstacles that throw up the huge waves, whirlpools (known as "holes") and strong currents that constitute a rapid. Crystal Rapids, 11 miles below Phantom Ranch, went from an easy rapid to one of the most difficult in the Canyon, after a flash flood in 1966 rolled rocks and boulders into that stretch of the river. The summer thunderstorms that can drop an amazing amount of water in a short time are also responsible for one of the prettiest sights to be had on a trip through the Grand Canyon—that of hundreds of miniature waterfalls streaming over the cliffs of the Canyon.

At mile 116 is Elves Chasm, an oasis where waterfalls drop into clear pools, and the mosses, ferns and other lush vegetation provide a striking contrast to the vast stretches of barren rock. At mile 134, Tapeats Creek roars its way down to the river. One of its tributaries, Thunder River, originates from an enormous spring which gushes out of a Redwall canyon. This is another oasis in the desert, full of the green of cottonwood and willow.

Yet another scenic wonder just downriver from Tapeats is Deer Creek Falls, which Powell floated by: "Just after dinner we pass a stream on the right, which leaps into the Colorado by a direct fall of more than 100 feet, forming a beautiful cascade. Around on the rocks in the cavelike chamber are set beautiful ferns, with delicate fronds and enameled stalks. This delicate foliage covers the rocks all about the fountain, and gives the chamber a great beauty."

Twenty miles more and the river traveler comes to Havasu Creek, named after the Havasupi Indians who live nearby in a village called Supai. Today there are more Havasupi than can fit in the small canyon valley they call home, but the 200 or so who live there cultivate fruits and vegetables and raise horses, existing largely in a world of their own. Havasu

Falls has probably inspired more photographs than any other single spot within the Grand Canyon. It is actually a series of streams that plunge 1,400 feet. The waterfalls, some tiny, some like Mooney dropping nearly 200 feet, fall into lovely pools made a clear turquoise blue by the limestone of the stream beds.

At mile 179 are the rapids that all boatmen have in the back of their minds as they descend the river, the last great obstacle of the Canyon, and one of the most awesome, difficult and dangerous stretches of white-water in the world, dropping 37 feet in 200 yards—Lava Falls. Powell, who wisely made a portage around Lava Falls, wrote, "What a conflict of water and fire there must have been here! Just imagine a river of molten rock running down into a river of melted snow. What a seething and boiling of the waters; what clouds of steam rolled into the heavens!" Geologists believe that at one time this lava flow completely blocked the river; today it is a stupendous obstacle course of gigantic boul-

ders which send waves 20 feet into the air when the water is high.

Three days after Powell and his men passed Lava Falls, three of his party left, having decided to hike out of the Canyon to one of the Mormon settlements they knew were above. Just two days later, Powell and the remaining members of his party emerged victorious from the Grand Canyon at Grand Wash Cliffs, close to where Lake Mead is today. The men who had hiked out were killed by Indians.

Taking a river trip through the Grand Canyon these days involves making reservations as much as a year ahead. To preserve the Canyon for everyone, the Park Service allows only a limited number of people through each summer, and demand is high. Two-week trips are made in oar-powered wooden dories, much like the ones Major Powell traveled in, or in rubber boats, some with outboard engines on them. The motorized boats make for a faster, but noisier trip down the river.

For decades after Powell, very few people attempted the journey through the Grand Canyon. Between 1869 and 1949, only 100 people ran the river. In the early 1960s as attention was focused on the Glen Canyon Dam and environmental awareness grew, the numbers grew as well, and by 1969 over 6,000 people had run the river in a year. Now the number has to be limited to about 15,000, and commercial outfitters who are on intimate terms with the river and its rapids take most people through.

There have been other changes as well. The wildlife Powell saw consisted mainly of deer, bighorn sheep, lizards, snakes and birds. Now wild burros, left behind by prospectors, overpopulate the Canyon, and the bighorn sheep, whose grazing territory the burros have all but taken over, are scarce. The changes in the water below the dam have made a fish called the humpback chub an endangered species, but they have made it more amenable to trout. The tamarisk that grows in dense clumps along the river is also a newcomer, introduced to America by the Department of Agriculture in the early 1900s in a misguided attempt to hold riverbanks in place and create windbreaks. It spread quickly and became a nuisance, as well as forcing out other riverbank growth such as willow and cottonwood. Nevertheless, the mosses, the ferns, the orchids and many other kinds of vegetation still grow in the Canyon, and seen simply as vegetation, even tamarisk can be a pleasure.

After he had made his trip through the Grand Canyon, Major Powell wrote: "You cannot see the Grand Canyon in one view, as if it were a changeless spectacle from which a curtain might be lifted, but to see it you have to toil from month to month through its labyrinths. It is a region more difficult to traverse than the Alps or the Himalayas, but if strength and courage are sufficient for the task, by a year's toil a concept of sublimity can be obtained never again to be equaled on the higher side of Paradise."

The wild variety of the **bighorn sheep**, also called the Rocky Mountain sheep, has been driven from its gazing land by the burro and is now in danger of becoming extinct.

Overleaf Every view of the Grand Canyon from above evokes a different mood and an inconceivably vast panorama of carved rock.

THE RIM

Exploring the rim of the Grand Canyon is a completely different experience from exploring the bottom. On the rim, the incredible red rocks stretch out to the horizon like a picture postcard, except for changes in light and shadow. Sunrise and sunset are the best times to see the Canyon from the top, because that is when the rapidly changing light is most dramatic. There are frequently afternoon thunderstorms during the summer that also make for exciting panoramas. The distance and enormity of the Grand Canyon from above is hard to comprehend, and one of the best ways to solve this problem it to hike down one of the trails that descend into the Canyon, or take a mule trip if trudging up and down on your own two feet doesn't appeal.

The piñon tree produces pine-nuts which formed a staple diet for the early Indians.

The South Rim

The South Rim is where most people begin, and it is guaranteed to be crowded with the thousands of others who have the same idea throughout the summer and early fall. Though it may seem hard to believe, having just driven through the scrub country of the Coconino Plateau to get there, the South Rim is at an elevation of 7,000 feet, and therefore ranks as mountain country, with temperatures that can get into the 90s during the day and drop to the 40s at night during the summer. Fall days on the rim are cool and crisp, and the first snows may fall by October.

Grand Canyon Village, on the South Rim, is open year-round and has tourist facilities including a museum, a nature trail, lodges, restaurants, a trailer park, campgrounds, a visitor center, post office, car rentals, bus tours and of course the ubiquitous souvenir shops. There are numerous viewpoints along the East and West Rim Drives, and trails that go a short distance down the Canyon. It is possible to hike to Havasu Canyon and Mooney Falls from the Topocoba Hilltop Trail, which begins on West Rim Drive.

Anyone who is over 12 years old, weighs under 200 pounds and is properly dressed and equipped can take a mule ride to the bottom of the canyon. These sure-footed beasts may appear to be half-asleep as they plod up and down the trail, but they've got a perfect safety record. You can either take a 12-mile round trip to the Tonto Plateau, about two-thirds of way down, or take two days and spend the night at Phantom Ranch. There are short and long day hikes all through Grand Canyon National Park. The most popular overnight hike is down the Bright Angel or Kaibab Trail to Phantom Ranch, and up the Kaibab Trail on the other side, to the North Rim. These are the only maintained trails to the bottom in the park. The Bright Angel Trail has three spots where water is available. The descent takes one through four life zones, from the Transition zone of piñon, Ponderosa pine and juniper forests on the rim, down to the Lower Sonoran zone of cacti and yucca. Along the river is yet another type of life zone predominated by willows and tamarisk. During the winter, when the North Rim is buried in snow, there may be flowers blooming far below in the bottom of the Canyon.

The North Rim

Though the North Rim is only ten miles from the South Rim as the crow flies, it is a 200-mile drive by car to get from one to the other. The distance between the two rims is vast in more ways than one. Two species of squirrel have evolved on the rim of the Grand Canyon, the Abert and the Kaibab. Though they are about the same size, which is large for squirrels, have the same habits, and the same basic color, their markings are completely different. The Abert has a white underside and a gray tail; the Kaibab has a black underside and a white tail. Geography has allowed the Abert to spread to other parts of the Southwest, but the Kaibab squirrel is found only in the 350 square miles of ponderosa forests on the Kaibab Plateau, marooned on an island separated from the world on one side by a mile-deep chasm, and on the other by deserts. There are evolutionary differences between other small

Both rims of the Grand Canyon provide many spectacular viewpoints. Most are accessible by car, but some can be reached only by hiking. **Opposite and above** the rims are 7,000 feet or more above sea level. The Canyon is about a mile deep; its width ranges from about 600 feet to 18 miles. Measured along the Colorado River, it is 277 miles long.

Left There are still a few spots in the West where one can see a herd of buffalo, now carefully watched over and protected.

Buffalo, who 200 years ago roamed across America in their millions, are now herded in protected reservations such as House Rock Valley Buffalo Range on the Kaibab Plateau.

mammals, such as mice, on the two rims, and in the Canyon itself there is a species of rattlesnake that has evolved to a pink color, to match the rocks in which it lives.

The North Rim is nearly 1,000 feet higher than the South Rim, and is closed for most of the winter by snow. This is the Kaibab Plateau, out of which the Canyon was carved, and here there are aspen groves, which turn a brilliant yellow in autumn, and spruce and fir trees. The North Rim is not usually as crowded as the South Rim, and has fewer commercial facilities. The views from the North Rim give a very different impression of the Grand Canyon than the South Rim. It is possible to hike in to Thunder River and Tapeats Creek from the Thunder River Trail, which begins at the North Rim.

Tuweep, located west of the North Rim offers the most isolated spot to view the Grand Canyon. Getting to it means driving over 65 miles of dirt roads with no commercial facilities at all. The wise fill their gas tanks and take plenty of water. Toroweap Point provides one of the most breathtaking views of the Canyon, with a sheer drop of 3,000 feet. Tuweep is above Lava Falls, so this is the spot to see the lava flows that at one time dammed up the Grand Canyon. A difficult trail descends here to Lava Falls.

On the way to or from Tuweep is Pipe Spring National Monument. This is a Mormon Fort built in the 1870s to protect Pipe Spring, one of the few watering holes in the area. The buffalo don't roam in America anymore, but at House Rock Valley Buffalo Range, west of Navaho Bridge on the Kaibab Plateau, is a herd of this nearly extinct species.

Tourists began flocking to the rim of the Grand Canyon as early as the 1880s. One of the first people to take advantage of this was Captain John Hance, who built a log cabin on the South Rim and mined copper and asbestos below. Putting his hard labor in carving a trail down the cliff walls to good use, Hance began guiding visitors to the floor of the Canyon. By the 1890s, Bright Angel Trail, Grandview, Tanner, Boucher, and Hermit had also been carved out, and there was even a ferry across the river. A stagecoach brought people in from Flagstaff, Arizona. In 1928, the National Park Service took over, leasing rights to concessionaires such as Fred Harvey, who had restaurants all over the Southwest. The Fred Harvey Company still operates the lodges at the South Rim.

Scraggly pine and juniper trees
cling to the sides of the Grand Canyon,
providing welcome patches
of shade in the summer.

The Kaibab Forest, on the North
Rim of the Grand Canyon, is a unique
ecological niche, isolated by
the Grand Canyon on one side and the desert
on the other.

The beauty of Oak Creek Canyon **opposite** has made nearby Sedona, Arizona, an artist's colony and popular tourist stop.

Phoenix **above** has become the capital of the Sunbelt, attracting millions of people in the past few decades. Many are wondering what effect an inevitable and permanent water shortage will have when Phoenix uses up its supply of groundwater.

The San Francisco Peaks **middle** are sacred to the Hopi and Navajo Indians, representing, among other things, a boundary of their homeland.

Tucson, Arizona, **below** is one of the booming Southwest cities whose existence depends on electricity and water from the Colorado River.

Flagstaff

Arizona has three large cities: Tucson in the south, Phoenix in the middle and Flagstaff in the north. Flagstaff is the smallest of the three, but it has as much or more cultural and geographical appeal as the others. In many ways, Flagstaff offers the best of both worlds—the warmth and sunshine of the desert only a few dozen miles away, and the cool summers and snow-filled winters of the mountains in which it is situated, 6,900 feet above sea level. In the winter, there is downhill skiing at Arizona Snow Bowl and cross-country skiing in the surrounding mountains. All the mountain sports are there in the summer, as well as the Pow Wow, a huge Indian celebration with dances, parades and rodeos, and the Summer Festival, two weeks of classical music, theater, films and art exhibits. Above Flagstaff, where the air is clear, the Lowell Observatory, nearly one hundred years old, provides an ideal spot to watch the stars. The Museum of Northern Arizona is practically a must for those touring the Southwest and the Grand Canyon. The displays on the flora, fauna, geology, anthropology and archaeology are superb, both attractive and informative.

To the south of Flagstaff, the highway descends rapidly through Oak Creek Canyon, a spectacular gorge that is second only to the Grand Canyon for scenic beauty in Arizona. Here the red-rock formations contrast with pine and aspen forests and the creek tumbling down the steep incline. At the foot of Oak Creek Canyon is Sedona, a small picturesque town full of artists and craftspeople attacted by the views in every direction.

Overleaf Sailors seek out the utter quiet and privacy found the endless canyons, bays and inlets of Lake Mead.

PART III
THE LOWER COLORADO

LAKE MEAD AND HOOVER DAM

The large, fleshy stems of **cactus plants** enable them to store water and survive desert droughts.

The monolithic white plug of the Hoover Dam **right** towering 726 feet above the river, is still awe-inspiring, as is the amount of electricity it generates for millions of people.

Though the shores of Lake Mead **opposite** are barren desert, the lake itself is full of bass and other game fish.

Lake Mead begins a few miles into the southern end of the Grand Canyon near Separation Rapids, named for the spot where three of Major Powell's party left the river expedition and climbed out of the Canyon, only to be killed by Indians. The lake is shaped somewhat like a swallow flying south, with the east wing the southern end of the Grand Canyon, the west wing nearly in Las Vegas and the forked tail pointing north into Nevada. One fork is the Virgin River; the other, Fork Meadow Valley Wash. The head points south into the desert and towards the Black Mountains. From wing tip to wing tip, the lake stretches for 108 miles, but with the long tail stretching north, and the many side canyons, there are more than 822 miles of shoreline.

Lake Mead was formed when the Hoover Dam, a great concrete plug in the Black Canyon, was built. The inspiration for building Hoover Dam grew out of the Colorado River Compact, which divided up Colorado River water between seven western states. This dam was the first major harness on the river,

which in the past had uncooperatively flooded every spring, breaking through the small earth dams built to hold it in and wreaking havoc on the crops, homes and roads built on its flood plain. By summer, the river would run nearly dry, making it useless for irrigation. Now, with a two-year supply of water backed up in Lake Mead, the flow of the water is controlled for the use of irrigation and hydroelectric power all the way to the Mexican border.

Construction on Hoover Dam began in 1931 and was finished by 1935, when it was dedicated by President Franklin D. Roosevelt. The first generator began churning out electricity in 1936. The massive arch gravity wedge of Hoover Dam rises 726 feet above the river, is 660 feet thick at its base, 45 feet thick at the top, and 1,244 feet long from canyon wall to canyon wall. The 4.4 million cubic yards of concrete poured to make the dam involved four years of round-the-clock work.

Once the dam was completed, at a cost of 400 million dollars, it quickly became apparent that Lake Mead was being filled with silt at a distressingly rapid rate. The river that had carried an estimated 180 million tons of silt to Mexico every year was now leaving about 165 tons of it behind the dam. In a lake with a potential capacity of storing 28.5 million acre-feet of water, 137,000 acre-feet a year were being displaced by silt, and 800,000 acre-feet were evaporating into the dry desert air every year. Simple arithmetic made it clear that if something wasn't done soon, Lake Mead would be a giant mud puddle before a century had passed. This was one of the major justifications for building Glen Canyon Dam—a "robbing Peter to pay Paul" logic that has now transferred most of the silt problem to Lake Powell.

Today, the 17 generators at Hoover Dam, fueled by the Colorado River, put out up to 1.34 million kilowatts of electricity. Southern California is the largest consumer of Hoover Dam-generated electricity, gobbling up more than 65 percent of the power generated. California also takes the lion's share of water out of the Colorado from Lake Mead south.

Most of Lake Mead's shoreline gives way to a barren, forbidding landscape of ancient lava flows, bare mountains and strange rock formations. Vegetation is limited to the Joshua trees and cacti that can survive in a climate which would seem more appropriate to the moon.

The top of Hoover Dam **above** is open to visitors. Good views of Lake Mead, which stretches 105 miles up the old course of the Colorado River, can be had from here.

The deep, starkly beautiful chasm of the Black Canyon **right** before it reaches Hoover and Davis dams is carved out of volcanic rock.

However, the three major marinas on the lake—Echo Bay, Temple Bar and Lake Mead—enjoy a year-round tourist business, thanks to a climate that tends toward balmy temperatures and blue skies, though it can be extremely hot in the summer and brisk in the winter. One of the most popular ways to explore the seemingly endless canyons of Lake Mead is by houseboat. These and sailboats, powerboats, fishing equipment, scuba-diving equipment and water-skiing equipment can all be rented at the marinas. The lake is well stocked with catfish, largemouth bass, rainbow trout and striped bass, all of which seem to thrive in the clear blue waters of the lake, so the fishing is excellent. The water level in Lake Mead fluctuates according to demand for water and snowmelt in the mountains to the north. Marina facilities were built that could accommodate different water levels after they were left high and dry on mud flats during a dry year.

Thousands of vacationers flee winter cold and snow in the nearby mountains to frolic on Lake Mead **above** under the warm desert sun.

Hoover Dam **left** has been churning out electricity for Western states since 1936.

Exploring the many-sided canyons
of Lake Mead is a popular pastime for
boaters.

The Hoover Dam was the first
large dam built on the Colorado River.

A bigtooth maple **opposite** lends a rare shade of brilliant red to an autumn in Utah.

Blowhead Point at Cedar Breaks National Monument in Utah **left** offers the contrasts of aspens, conifers and carved red rock.

Zion Canyon

The Virgin River, a major tributary of the Colorado, has its headwaters at two forks, both originating in the area of Cedar Breaks National Monument in southwest Utah, on the Markagunt Plateau. The North Fork has helped carve a 2,000-foot-deep canyon through the shale and Navajo sandstone that form the White and Vermillion Cliffs above the river. Zion Canyon, its colorful rock walls and intricate, narrow side canyons contrasting with the green valley below, is one of the most magnificent national parks in the country, and well worth a detour from the Colorado Plateau. The Indians avoided settling in Zion Canyon, believing it to be sacred. Mormon settlers took advantage of this fact and established farming communities there until the National Park Service took it over in the 1920s. There are still Mormon farming communities in the lovely green valleys along the river south of Zion National Park.

Las Vegas

The glittering neon of Las Vegas is an anomaly in the Southwest, an oasis in the middle of the desert, watered by the Colorado River and lit up by the Hoover Dam. The city owes its origins to springs and meadows that were probably frequented by the Paiute Indians, and named Las Vegas (the meadows) by Spanish explorers passing through in the late 1700s. Traders, trappers and wagon trains using the Spanish Trail to get to California later took advantage of the rare patch of grass and water. In a pattern typical of this area of the country, the Mormons were the first to settle there, to be followed by cattle ranchers. The settlement quickly became a boom town for a few years when gold and silver were discovered nearby in 1849, then faded into obscurity again until the early 1900s when the Union Pacific Railroad was built through it. Even with a railroad, there wasn't much to support the town, and it remained a quiet

The **golden eagle**, found in open mountains, canyons and plains, glides through the air with only an occasional wingbeat before swooping to catch its prey of rabbits and larger rodents.

Overleaf The Zion Watchman is one of many spectacular rock formations in Utah's Zion National Park.

Right The red sandstone and shale of the Southwest become coral pink sand dunes when they erode.

depot until construction on the Hoover Dam began in 1931 and Nevada legalized gambling.

But for the gambling, Las Vegas might have faded again once the dam was built. However, the chance to win (or lose) big proved to be an irresistible magnet, and today 12 million people a year (and rising) try their luck at the gaming tables and slot machines of Las Vegas. If there is a way to lure customers, it has been thought of in Las Vegas—gargantuan neon signs which are the city's hallmark; building themes which include the circus big top, grandiose palaces and tropical jungles; entertainers who do everything from sing and dance to magic tricks and strip tease; brothels; and, of course, the quickie divorces and

marriages pioneered by Las Vegas. Lately, the city has been attracting major sporting events, and the convention business has long been a booming one.

"The Strip" is famous for its monumental hotels that a person could enter and need not leave for weeks—or, as the reclusive millionaire Howard Hughes proved, for years. Some of the biggies are Caesar's, the MGM Grand, the Sands, Circus-Circus, the Barbary Coast, the Flamingo Hilton, the Dunes and the Imperial Palace. Downtown, in "glitter gulch," smaller gambling casinos and souvenir shops crowd the streets, and what they lack in size, compared to the rambling "uptown" hotels, they make up for in quantity and sheer,

Indian Paintbrush is a hardy
and colorful plant found throughout the West.
Here it is dwarfed but not outdone
by a canyon wall in Zion National Park in
Utah.

unashamed, unadulterated brassiness.

Las Vegas still has a fairly large Mormon population, and there are many government workers who live there, employed by Nellis Air Force Base and the Atomic Energy Commission's Nevada Test Site, located north of the city. Las Vegas is also seeing growth in other kinds of industry, primarily warehousing and distribution centers, attracted by the lowest tax rates in the country.

Las Vegas gets about half its water from Lake Mead, and half from groundwater. Unfortunately, groundwater is not necessarily replenishable, and the fountains, golf courses, homes and hotels of Las Vegas have depleted the supply to such an extent that the land under the city has sunk more than three feet. Las Vegas faces the same problems of other oases in the Southwestern desert, such as Phoenix and Tucson, with a burgeoning population and a diminishing supply of groundwater. Both Las Vegas and Phoenix are scheduled to begin taking more water out of the Colorado River in the mid-1980s, events that will precipitate a crisis that has been pending since the Colorado River Compact apportioned more water than was available. There simply isn't enough to go around, and nobody is willing to give up a drop. It looks like the Colorado River will continue to be the most litigated, dammed and diverted river in the world for some time to come.

Lake Mohave

Below the Hoover Dam, the Colorado River turns south, forming the border between Arizona and California. Its flow is backed up in reservoirs for such of its remaining length. The first of these is Lake Mohave, formed by the Davis Dan, 66 miles south of the Hoover Dam. The upper end of Lake Mohave covers yet another spectacular canyon which has gone to a watery grave. The Black Canyon was sliced out of volcanic rock, and the remaining cliffs rise nearly straight up, hemming in most of the northern end of the lake. The Davis Dam was completed in 1951 to provide Mexico with its share of Colorado River water.

Like Lake Powell and Lake Mead, Lake Mohave **above left and right** creates a year-round playground in the sun for water sports, yet is surrounded by desert. The lake is created by Hoover Dam to the north and Davis Dam to the south.

When the Mexican government protested that the Colorado River no longer reached the border, the Davis Dam **below** was built to create a reservoir for Mexico's use.

LIFE IN THE DESERT

Seen from a distance, the desert may appear to be lifeless and barren, or an alien environment supporting only a few kinds of life. A close-up view reveals that a unique ecosystem survives in the desert, abounding with plants and animals that have adapted to a climate where rain may not fall for more than a year and temperatures regularly top 100 degrees Farenheit. From Lake Mead south, the Colorado is surrounded by the Mohave Desert to the west and the Sonoran Desert to the east. Blythe, Needles and Yuma regularly record the hottest temperatures in the nation during the summer. If the Colorado was flowing in its natural state, in some years the water would be low enough to wade across. These desert climates are created by weather patterns that create large and relatively stationary masses of dry, high-pressure air called semipermanent subtropical anticyclones. Anticyclones occur between the latitudes of 20 and 30 degrees north and south of the equator, creating a ring of deserts around the earth that includes the Australian deserts, the Sahara, the Gobi and about a dozen others.

Different areas of these deserts support different kinds of plant life. The Joshua tree with its dark, bristling branches is seen in the Mohave; the magnificent saguaro cactus with its tall trunk and upstretched arms and the thin branches of the ubiquitous and hardy creosote bush are seen in the more mountainous Sonoran Desert. There are at least a hundred kinds of cactus growing in the Sonoran Desert, including the senita, prickly pear, mound cactus, barrel cactus and beavertail cactus.

When it does rain in the desert, the water rushes down the slightest incline, creating arroyos (or gashes) in the desert floor. Flash floods cut the arroyos deeper and deeper, and some water is left behind to seep into the ground. Desert growth such as mesquite, ironwood and palo verde tend to congregate in or near these arroyos. This same action of water tumbles rocks and other detritus down mountainsides, creating a kind of delta that spreads out from the base. This is where the largest concentration of cacti, creosote bush, cholla and yucca tend to congregate.

Every desert plant has its own system for conserving water. Some have very deep taproots which may even find groundwater, others have shallow root systems that spread out to soak up every available drop of moisture. The cactus has spines which not only protect it from predators but also lose far less moisture than leaves would. Beneath the tough skin, which takes over the process of photosynthesis that leaves normally perform, literally tons of water can be soaked up into a large saguaro cactus during one rainfall. The longer the next drought lasts, the deeper the pleated ribs of the saguaro get, as it uses up its precious store of water. The saguaro also grows tiny hair-like roots immediately after a rainfall. These roots help in collecting moisture and then shrivel up until they are needed again. Desert plants also tend to space themselves far enough apart so that each one can survive on the surrounding water which comes its way. In a prolonged drought, which in the desert may mean more than a year, many plants go into what amounts to a state of hibernation.

The spiky **Joshua tree**, a species of Yucca native to the Mohave Desert, was given its name by Mormons who likened its branches to arms stretched out to heaven.

Desert hikers **right** should carry lots of water and beware of the many spiny desert inhabitants like this Spanish bayonet, a type of yucca.

The upraised arms and striking silhouette of the giant saguaro cactus **opposite** have become a symbol of the Southwestern desert.

D E S E R

P L A N T S

As if to make up for
their astonishingly brief lives,
frail desert flowers,
their seeds lying dormant for
most of the year, come
in bright colors and exotic
shapes. Shown here,
clockwise from the top left,
are the ocotillo, the
pincushion cactus flower, the
calico cactus flower,
the desert mariposa and the
prickly pear flower.

The ocotillo survives in the desert
by minimizing its foliage; small, fleshy leaves
appear after rain, and soon drop.
Red flowers sprout and drop just as quickly
after summer showers.

Organ Pipe National Monument
in Arizona is the only place in the world
where the organ pipe cactus
is found.

The desert is a positively joyous place to be after a rainfall. Within hours, plants begin to bloom or grow leaves. The aroma of mesquite, palo verde and ironwood fills air, and suddenly there are insects buzzing about. In a matter of days or weeks, all this abundance disappears again until the next rainfall.

The animals of the desert have also made unique adaptations to drought and heat. Most of them only come out at night when the temperature drops—in a desert that is over 100 degrees during the day, night-time temperatures can be downright brisk. Larger mammals such as the mule deer, peccary, bobcat, coyote, jackrabbit, kit fox, gray fox and ringtailed cat depend on the waterholes they visit at night. Most of these are shaded rock-basins (called tinajas) which can catch enough water to last for months. Sometimes there is groundwater just below the surface that a savvy animal can dig down to. Or in the Sonoran Desert, where rain falls on the mountains, they may simply go uphill to find water.

The kangaroo rat, with its long hind legs and long tail, is by far the best example of a species that has learned to do without water, for they have literally done that—in fact, they

The giant saguaro cactus **right** has a very limited habitat in Arizona. The huge trunks are capable of storing hundreds of gallons of water.

Previous page Many varieties of yucca grow in the desert, protected from predators by their sword-like spines. The Indians used the liquid from mashed yucca roots as a shampoo.

Above Creosote bush, ironwood, mesquite and salt bush are a few of the hardy shrubs that grow in the desert. Willows grow near the rare stream or spring.

As if to make up for lack of water and thus foliage, desert plants **left** take on fantastic shapes.

The **kangaroo rat**, which gets all the water it needs from its food, inhabits arid desertland and uses its powerful, long hindlegs to move by swift leaps.

won't drink even if offered water—they don't know what to do with it! The seeds and grasses eaten by kangaroo rats contain hydrogen that combines with the oxygen they breath to produce water internally. But that's not all. Their nostrils are cooler than the rest of the body, so warm exhaled air condenses and returns to the system. Their kidneys are so large that virtually no moisture escapes in their urine or feces. They plug their burrows, which they sleep in during the day, with sand, so any vapor that does escape through exhalation can be reabsorbed. When food is plentiful, kangaroo rats breed prodigiously and provide food for larger predators.

Other rodents, such as the pack rat, pocket mouse, ground squirrel and wood rat may find moisture in cactus pads or the stems, leaves and roots of plants. They have evolved their own niches so neatly that while one species lives on the stems of a plant, another lives on the roots.

The snakes, lizards and toads that inhabit the desert get much of their moisture from their prey, whether small mammals or insects. They too usually burrow during the day and come out at night, though lizards can often be seen at dusk and dawn hunting for insects. The desert tortoise can conserve large amounts of water under its shell, and burrows into the ground when it gets too hot. The hard shell also protects them from predators. One species of desert toad and some snakes actually hibernate through the driest and hottest months, or through the winter in the colder parts of the desert. The tadpole shrimp is found in waterholes, but when they dry up, the small crustacean burrows deep into the mud, staying dormant for years sometimes until a sufficient rainfall wakes it up again. Scorpions are also night creatures. Anyone planning to prowl about the desert at night is wise to wear high boots to protect from snakes, the painful sting of the scorpion, or even the bite of that big, hairy spider the tarantula, which is normally shy and retiring.

Birds of the desert can range from white-winged doves, thrashers, orioles, blackbirds, warblers, bluebirds, flycatchers and sparrows to bats, hawks, elf owls and crows. Most desert birds live happily amongst the spines of the cactus, protected from predators, eating its fruit and sipping the nectar of the large showy blossoms.

The desert, despite its barren surface appearance, teems with life. Crea of all kinds adapt to its harsh conditions — lizards who shelter from heat of the day under rocks, armor-plated scorpions whose bite is pa and even the delicate and seemingl fragile Mountain Emperor butterfly.

NAVIGATING THE COLORADO

When the Spanish had conquered the Aztecs and taken over Mexico, they began exploring to the north in earnest, lured by tales of cities made out of gold, the Seven Cities of Cibola. While some parties explored by land, others were sent out to explore by ship. In 1539, the Spanish Viceroy Don Antonio de Mendoza put Francisco de Ulloa in charge of three ships that were to sail up the coast of Mexico and look for an easier overland route to the Seven Cities of Cibola than had so far been found in the inhospitable deserts. Ulloa was the first to record a run-in with the dreaded tidal bores of the Colorado River, which have sunk many a ship since. He wrote, "We perceived the sea to run with so great a rage into the land that it was a thing to be marveled at; and with a like fury it returned back again with the ebb. ... There were diverse opinions amongst us, and some thought that some great river might be the cause thereof." However, Ulloa did not try to fight the tidal bores, and returned without seeing the river or finding the legendary cities.

Viceroy Mendoza was not to be deterred, and a year later organized an expedition that would be coordinated between land and sea. Hernando de Alarcon was put in charge of three ships, the *San Pedro,* the *Santa Catalina* and the *San Gabriel.* He was to sail north into the Gulf of California—then called the Sea of Cortez—and meet the overland party which was led by Francisco Vasquez de Coronado. Upon reaching the head of the gulf, the three ships were immediately beached on sand bars by the particularly fierce tidal bores which can occur in late summer: "Whereupon we were in such jeopardy that the deck of the *Admiral* was oftentimes under water; and if a great surge of the sea had not come and driven our ship right up and gave her leave, as it were, to breathe awhile, we had there been drowned."

Perhaps Alarcon was more foolish or more courageous than Ulloa, or perhaps he had been threatened with dire consequences if he failed to reach the overland party. In any case, he persisted and said: "Now it pleased God upon the return of the flood that the ships came on float, and so we went forward. And we passed forward with much ado, turning our sterns now this way, now that way, to seek and find the channel. And it pleased God that after this sort we came to the very bottom of the bay, where we found a very mighty river,

which ran with so great fury of a storm, that we could hardly sail against it."

Alarcon did manage to get up the Colorado to a point about a hundred miles north of the Gila River, but not without the aid of the Cocopah Indians he had somehow talked into towing his ships up the river. Not finding the Coronado party, Alarcon left letters for them in a cache under a tree and returned home.

Three hundred years later, in 1829, a British lieutenant named Robert W.H. Hardy was sent to the west coast of Mexico to explore rumors of rich oyster beds. He managed to navigate up the Colorado to its confluence with the Gila, providing the first accurate charts.

The Steamboat Eva

Once Fort Yuma was built, around 1851, a man named George A. Johnson began a steamboat company which ferried supplies up from the Gulf of Yuma. His boats, eight in all, were small stern-wheel steamers with shallow bottoms, and the business lasted until after the Civil War. The next person to try and navigate up the river was the same Lieutenant Joseph Ives who later called the Grand Canyon a "profitless locality." The War Department built him a 58-foot steel-hulled steamboat called the *Explorer,* much too big and heavy to make the journey anything but tedious, which it was. Thanks to long hours and days spent towing the boat through rapids and off sandbars, it took Ives and his men a month to get 150 miles north of Yuma. They struggled north to Black Canyon through rapids and over rocks for another month before Ives sent the *Explorer* back and continued on to the Grand Canyon.

About ten years later, the eminently foolish Samuel Adams, the same man who, a few years later, tried to descend the Colorado from the Blue River, made a similarly ill-fated trip from the Gulf of California to Boulder Canyon. It was John Wesley Powell's trip through the Grand Canyon that finally convinced the United States government that the Colorado River would not serve as a thoroughfare comparable with the Mississippi.

Powell and his crew run the **mighty rapids** of the upper Colorado, the most treacherous stretch of the river.

Opposite Every year thousands of people float down the Colorado River from Moab in canoes, **148** kayaks and dories.

LAKE HAVASU

Lake Havasu is the stretch of Colorado River between the Davis Dam and the Parker Dam. The Fort Mohave Indian Reservation below Davis Dam straddles the Colorado, and was named not only after the Indians who inhabited it, but also after the fort built after the Mohaves were defeated by the United States Army in what was called a "military engagement"—a series of skirmishes in which the Mohaves never really had much of a chance.

The reservation was established after the battle. Nevertheless, the Mohaves were known among other tribes as fierce fighters, especially in hand-to-hand combat, and attacked many a prospector looking for gold in the late 1800s, causing many to give up and return home empty-handed.

The Mohaves are among a group of Indians that speak the Yuman language and traditionally lived along the Colorado in western Arizona, farming the soil that was made rich by the annual flooding of the river. Other Yuman tribes include the Havasupi, Hualapai, Yuma, Maricopa, Yavapai, Cocopah and Chemuevi. The Colorado Indian Reservation to the south also straddles the river and was originally set aside for the Mohaves, with the stipulation that Indians from other tribes could live there, too. When it looked like there might eventually be more Hopi on the reservation than Mohave, the tribe protested, and since 1957 it has been the domain of the Mohave only. The Hopi who moved there to farm have become largely acculturated to the Mohave way of life.

The Mohave and most of the other western Arizona tribes have always been agrarian cultures and continue to be so. To the increasing dismay of others who have a claim on Colorado River water, the Winters Doctrine of 1908 stated that the Indians whose reservations were along the river could use as much water as they neded. This was amended in 1963 to mean that they could use as much water as they needed for the purpose of irrigating their land.

If those rights are exercised, which the Mohaves are in the process of doing, a very large chunk of water could go to the Fort Mohave and Colorado Indian Reservations—yet another possibility not allowed for in the Colorado River Compact. Much of the Indian land is now leased out to non-Indians for farming purposes.

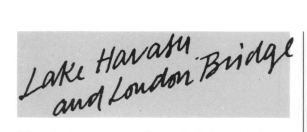

Lake Havasu and London Bridge

The forty or so miles of Colorado River between the Davis Dam and the Parker Dam to the south is called Lake Havasu, named after the Indian word for "blue water." Yet another narrow, scenic canyon accessible only by boat and named Topock Gorge leads into the lake. The buzz of powerboat engines in the gorge and on the narrow lake is constant. Only three miles across at its widest point, Lake Havasu is a haven for water skiers because for most of the year not much chop gets kicked up in the relatively small space. Lake Havasu City was founded and developed by Robert McCulloch, a chain-saw magnate who saw in it an opportunity to test and promote outboard engines as well as sell subdivision lots to Southwest emigrés seeking the sun and warmth of the desert and the recreational opportunities of Lake Havasu.

McCulloch put Lake Havasu City on the map when he purchased London Bridge for 2,460,000 dollars from the city of London. For another 5.6 million dollars he had it dismantled, shipped 10,000 miles, and rebuilt over a core of steel and concrete when it now spans the Colorado River. Actually, the bridge was built on dry land and the river was then diverted to flow under it. On a river that amounts to no more than a controlled ditch at this point, this particular feat of engineering and public relations seems appropriate. Whether the London Bridge and the surrounding desert grace each other is a matter of debate, but there is no question that it is an anomalous landmark and a target for superlatives, such as the largest antique ever sold. For those travelers made nostalgic for a pint and some fish and chips, there is an English village with a pub, restaurant and shops at one end of the bridge.

The Bill Williams River, which empties into the Colorado from the east just above Parker Dam, is a short, steep desert river named after the legendary mountain man. The bulk of the water contributed by the Bill Williams comes in the form of flash floods, but even that is precious in the desert, and the Alamo Dam was

Yuma dolls, male and female, wear the traditional costume of the tribe.

Parker Dam **opposite** backs up Lake Havasu, where thousands of "snowbirds" fleeing cold winters in the north camp for the winter.

I N D I A N

Many Indian tribes once farmed the rich flood plains of the Colorado River in present-day Arizona and California. Pottery of the Colorado River Indians is sought for its simple, graceful lines.

Baby carriers served many functions, from providing shade and protecting the head, to a vehicle for talismans.

POTTERY

Petroglyphs **above** are one of the few remaining traces of the Indians' culture.

A section of Lake Havasu City **above** and **below** is built to look like—you guessed it—London, complete with pubs and bric-a-brac shops.

Overleaf The clouds that hover over the mountains outside of Lake Havasu City rarely reach the desert.

built on it to capture the irregular torrents.

Parker Dam was completed in 1939, built by the Bureau of Reclamation and the Metropolitan Water District of California. Its main purpose is to get water to southern California. Half the energy generated by the dam is used to pump the water 242 miles west via the Colorado River Aqueduct.

Yet another big ditch is soon going to be drawing enormous amounts of water out of the Colorado River at the Parker Dam. Ever since the Colorado Compact was drawn up in 1928, Arizona, claiming that California gets too much of the water, has been suing, all the way up to the Supreme Court—and losing. When California was allocated the bulk of power coming from Hoover Dam, the Parker Dam, built shortly after, was the last straw, and the governor of Arizona called out the National Guard, ordering them not to allow any con-

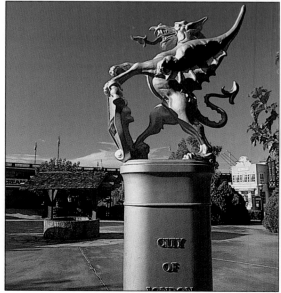

Who else but an American would
think to transport London Bridge to an
American desert to promote
a city? This is the Real McCoy.

struction to take place on the Arizona side of the Parker Dam. The ploy worked, up to a point. Construction was delayed for months and some concessions were won. But it wasn't until 1964, with a burgeoning population and more power in Washington, that Arizona really won the battle for water. Some of California's water is now due to go east instead of west to Arizona swimming pools, agriculture and copper companies and the Indian reservations in central Arizona. The water will get there through the Central Arizona Project (CAP), which will siphon 90 tons of water a second 815 feet up a tunnel through Buckskin Mountain and then through a series of pipes and canals for 310 miles. Meanwhile, the Metropolitan Water District in southern California hasn't figured out what it is going to do for thirsty farmlands and city dwellers when the CAP is completed. Nor have Phoenix and Tucson figured out what they are going to do when, in spite of the added boost from the CAP, their groundwater runs out, which it is projected to do in a matter of decades. The only untapped water left is in the Northwest, and although government engineers have cooked up many elaborate schemes to get Columbia River water south, none of them, so far, are feasible.

Gems and Ghost Towns

Below the Parker Dam is the Colorado Indian Reservation, and south of that are two very different but interesting towns, Blythe and Quartzsite. Quartzsite is actually 20 miles from the Colorado River, separated from it by the Dome Rock Mountains, but the town represents a phenomenon happening the length of the southern Colorado River—the migration of the Snow Belt set.

During the summer, Quartzsite, Arizona, is a sleepy desert crossroads near the California border with a few hundred residents who mainly try to stay out of the desert heat. As the days grow shorter and cooler, the population swells—up to 750,000 gray-haired Snow Belt

Lake Havasu **above** is only three miles wide, a haven for water skiers and power boats. Havasu is an Indian word meaning "blue water."

The desert is especially beautiful when it's in bloom, as here near the Parker Dam **below**.

Below When homes are located down long and often impassable dirt roads, mailboxes tend to congregate.

refugees and retirees from all over North America arrive in campers of every description. The main street becomes a winter-long flea market, and many of the seasonal residents spend their days in the desert collecting rocks. In February, they hold a "rock festival" called the Pow Wow, which features rough gems and minerals collected over the winter. It now draws over a million other visitors every year.

The population flocking to the Southwest every winter is particularly concentrated along the banks of the Colorado from Las Vegas south—there are literally hundreds of encampments. In Nevada, the gambling casinos quickly learned to accommodate the visitors with drive-up casinos and huge paved parking lots with camper cookups.

For decades, Quartzsite wasn't much more than a watering hole for prospectors, United States Army troops and, later, cross-country travelers. One of the most popular landmarks in Quartzsite is the grave of Hadji Ali, otherwise known as Hi Jolly, who came to Arizona from Syria in the 1850s with a herd of camels, at the request of the United States Army. The Army had decided to experiment with using camels for transportation in the Southwest deserts, and Hi Jolly was the imported expert, the chief camel driver. Among other things, the enlisted men didn't like the camel's habit of biting whatever part of the human anatomy happened to be within reach, and in less than a decade the experiment was scrapped. Most of the camels were auctioned off to zoos and the like, but those that didn't sell were turned loose in the desert. Hi Jolly kept a few and for a while operated a transport business between the Colorado River and nearby mining camps, but that didn't work either, and he turned his camels loose, too. Nobody really knows the fate of the camels, but most of them apparently turned rogue and were probably shot. For a few decades, stories filtered in of camel sightings in the desert, but by now they all seem to have died out, and camels do not seem destined to be used as beasts of burden in North America.

Blythe is in the Palo Verde Valley, which runs along the Colorado for about 35 miles and is the center of a rich agricultural and cattle-raising district. The Palo Verde Dam in Blythe was built not to store water but to apportion up to 1,800 cubic feet of water a

second to 300 miles of canals and irrigation ditches, one of the oldest irrigation systems on the river. With a 12-month growing season, the valley generates about 25 million dollars anually from crops such as wheat, cotton, melons and alfalfa to feed the cattle, and, in the winter, lettuce, much of which will end up in salads being tossed in the northeastern United States. The processing plants here epitomize agribusiness and are among the most tech-

nologically sophisticated in the world. They should be seen by anyone who wonders how a head of lettuce or cantaloupe can travel across a continent and stay fresh and intact.

There are signs of the ancient Indians who inhabited the area all over the Palo Verde Valley, mainly in the form of pictographs and petroglyphs. Just north of Blythe, on top of a mesa, are the remnants of giant petroglyphs etched out of the desert. The largest of them

appears to represent a female 170 feet long with outstretched arms 158 feet across, and quartzite stones marking her facial features and breasts. These mysterious figures are powerfully evocative, and though best seen from the air, they can be partially viewed from selected spots along maintained trails. Since two of the animals represented seem to be horses, scientists assume that the figures were made either 10,000 years ago when the Pleistocene horse inhabited the area or after the Spanish arrived with their horses in the mid-1500s. Dating methods would suggest the latter theory, but the Indians who live in the area claim to know nothing about the figures.

The stretch of river between the Palo Verde Dam and the Imperial Dam 90 miles south is the last truly navigable and pleasant stretch of the great Colorado River, bounded on both sides by the Imperial National Wildlife Refuge. The biggest attraction of this refuge is the bird life. The water runs lazily here, allowing protective plant growth along the banks. Some of the birds that might be spotted in the Imperial Wildlife Refuge are ducks, geese, heron, ibis, cormorants, quail, hawks, owls and doves.

A canoe trip down this part of the Colorado can be very rewarding, but it should be avoided in the spring by those seeking peace and quiet. In early spring, the Blythe River Cruise takes place, a powerboat tour from Blythe to Martinez Lake that attracts hundreds of participants. It may be a great treat for those under engine power, but kayaks and canoes can't compete with the noise, speed and chop created by the wake of all those boats. For those who enjoy powerboating in large groups, the tour is led by experienced pilots who know the pitfalls of the river.

The **Harlan's hawk**, whose coloring makes it easily confused with the red-tail hawk, is a casual visitor to Colorado and the southwest.

Oasis in the Desert

One of the great wonders of the desert is the ability of lush plants to take hold, forming an oasis wherever there is a reasonably steady supply of water. One of these is Corn Springs, west of Blythe, a verdant canyon of lush

Right Where irrigation
stops, the desert begins.
The canals crisscrossing
the Imperial make it the
richest agricultural valley in
the world.

tropical growth and Indian petroglyphs on the rock walls.

Another oasis called Palm Canyon is located in the Kofa National Wildlife Refuge southeast of Blythe. The Kofa Mountains are typical of the desert landscape here, with jagged, barren crags and pinnacles that support only a few kinds of hardy plant life.

The sudden appearance of a narrow canyon full of palm trees is a delightful contrast. Thousands of years ago, this species of palm was widespread in parts of Arizona, but with the exception of a few spots like this, most of them died off when the climate changed. There are also bighorn sheep in the Kofas, but it takes time and patience—or luck—to spot these shy

animals,

Southwestern Arizona is a haven for ghost-town aficionados. Dozens of mines were dug out of the hills, and in the short time that they produced precious metals, they also spawned small, temporary communities. The ruins of many have been preserved in the hot, dry desert air. The Kofas are a geographer's abbreviation for King of Arizona, the name of a rich gold and silver mine discovered there in 1896 by prospector Charles Eichelberger. Before the surface ores played out in 1911, the mine produced 3.5 million dollars worth of silver and gold. For seekers of tall tales and legends of buried treasure and lost mines, there are dozens in this desert.

MAKING
THE DESERT BLOOM

As early as 1901, various people had noted, with a gleam in their eye, that the riverbed of the Colorado River was higher than the desert to the west. It was clear that the river had flooded into this area before, because though it was desert, the soil was rich alluvial silt that had been deposited by the river—a farming bonanza if it could be irrigated. The first attempt at irrigating the Imperial Valley resulted in a disaster that took many years and millions of dollars to repair, as well as spurring the construction of the Hoover, Parker and Davis Dams. Geologists think that the Salton Basin, a depression below sea level just north of the Mexican border in California, was once the northern tip of the Gulf of California. The theory goes that as silt from the Colorado accumulated it formed a dam, and the river eventually took another course around it to the east. This is its present course.

When the California Development Company started selling land in the Imperial Valley in 1901 on the basis of the dams and canals it was building, people moved in by the thousands, settling in the harsh desert, battling rattlesnakes and dust storms from crude tents and shacks and waiting for the promised water. Land was cheap; it was the water they would really pay for. Water did begin to trickle into the valley, and though there still wasn't enough for everyone there, it was clear that anything would grow—and grow well—with the favorable climate, the rich soil and Colorado River water.

The California Development Company was in such a hurry to get the water to the Imperial Valley that they built their dams, headgates and canals quickly and carelessly, not stopping to think what would happen if the river swelled, which it did in 1905. The Colorado River literally changed its course and poured into the Imperial Valley, easily dissolving the crude dirt canals, washing away crops, homes, railroad tracks and anything else in its path. The Salton Sea began to fill up again, inundating the New Liverpool Salt Works. The California Development Company, which was

The Imperial dam was built to irrigate the desert west of the Colorado River **right** and **opposite**, a floodplain rich in alluvial soil deposits.

Overleaf Colorado River water irrigates vegetables that are shipped all over the United States from the
Imperial Valley.

Right An unusually blue and green stretch of the Colorado near Needles, California.

bankrupt on paper to begin with, could do nothing to stop the flow. Finally the Southern Pacific Railroad moved in, and, two years later, managed to get the river back under control. When the railroad refused to control a 1910 flood, the clamor began for the federal government to move in and build a secure headgate and canal. The Bureau of Reclamation complied, and began by building Hoover dam, creating another land rush in the Imperial Valley. The Imperial Dam and the All-American Canal were built to divert the water, and the Imperial Valley alone eventually ended up receiving more Colorado River water than any of the states in the Colorado River Compact, and 80 percent of the water allocated to California. This amounts to 450,000 acres irrigated by 3,000 miles of canals, most of them draining into the brackish Salton Sea.

In an effort to stop the monopolization of land and water in the West, when the Bureau of Reclamation took over the salvage of the Imperial Valley, it decreed that nobody could farm more than 160 acres. That was not only disputed, it was blatantly ignored. Today, the Imperial, the most productive agricultural valley in the world, is controlled by agribusiness, with few farms under 1,000 acres. It is truly a feudal valley, with underpaid migrant laborers—many of them illegal immigrants from Mexico—providing cheap labor and a few corporations raking in huge profits. After decades of battle, the United States Supreme Court finally made this legal, giving the land barons their way.

yuma

Yuma owes its existence to a river-crossing that brought thousands of travelers and prospectors through the area. It didn't take long for ambitious settlers to open businesses to provide for the travelers and ferry-boat services to get people across the river. This too is agricultural and cattle country, with nearly 300,000 acres of irrigated land. East of Yuma is the Yuma Proving Ground and Luke Air Force Base, which extend almost to central Arizona.

Overleaf The Glamis sand dunes, not far from Yuma, are one of the most popular attractions for visitors to the city.

The Glamis Sand Dunes near
Yuma are known locally as the 'Little Sahara.'
Dune buggy enthusiasts are
their most frequent visitors, followed by
photographers who catch shifting
moods of light and shadow.

The United States military has claimed many of the desert areas in the Southwest for war games and testing sites, so these are not places for exploring.

Centuries ago, the Quechan Indians, ancestors of the Yumans, irrigated the land around Yuma with Colorado River water, as did the Anasazi and Hohokam to the east. This was probably accomplished with crude canals and brush dams that were expected to wash away each spring with flood waters that also fertilized their fields with alluvial deposits. The Fort Yuman Indian Reservation is located on the west side of the Colorado.

Cabeza Prieta National Wildlife Refuge and Organ Pipe Cactus National Monument are located on the Mexican border, in remote and sparsely settled desert. Within Organ Pipe are the only cacti of that name on the United States side of the border. In the spring, the area is carpeted in a brief and riotous display of wildflowers.

At one time, the Gila River was a major tributary of the Colorado, coming in from the east after a long journey from the Mogollon Range in New Mexico. Until the Gadsen Purchase in 1853, the Gila was the border between the United States and Mexico. Today, nearly all of the Gila's water is diverted for irrigation, and its streambed is dry for most of the year.

West of Yuma are the rolling Glamis sand dunes known locally as "the little Sahara of America." What used to be an unpleasant obstacle faced by travelers leaving Yuma for California is today a haven for dune buggies. There are still remnants of an old plank road built so that cars could get across the dunes.

For its last 80 miles below Yuma, the Colorado River is a mere trickle, winding through a delta of marshes, mud flats and dry lakes, rarely if ever reacing the Gulf of California. For 27 of those miles it is "backed up" behind the Morelos Dam in Mexico, which diverts what is left back across the border to the Mexicali Valley. A complex system of canals south of Yuma divert waste water from irrigation in a belated attempt to give Mexico its share of water.

Yuma is the site of a 350-million-dollar desalinization plant being built by the Bureau of Reclamation. As the water which has irrigated millions of acres flows back into the Colorado, it takes silt with it that is heavily

saline. By the time this water gets to Mexico, the salinity is so high that it kills crops. Mexico protested, and the answer was this plant, which will use a sophisticated reverse osmosis procedure to filter the water before it goes to irrigate Mexican crops. But even the desalinization plant is incapable of delivering to Mexico its allotted amount of Colorado River water, so the United States government went south of Yuma and drilled 600-foot wells that pump water across the border. Mexico has drilled its own wells, tapping the same groundwater source, so it is just a matter of time before the supply is completely exhausted.

Left The Colorado River rapidly diminishes in size as it heads south from Yuma. For most of the year it's a muddy trickle even here.

173

INDEX

Picture Credits

Ellis Armstrong: pp.6, 12/13 (top), 14, 17, 18/19 (right),28/29 (top left, top right), 30/31 (top, left, right), 36/37 (top, left), 41, 42 (right),48, 78, 92 (left)

Colorado National Park Service: pp.32/33

Denver and Colorado Convention and Visitors Bureau: pp.12/13 (centre), 18/19 (left), 20, 21, 22/23

Tom Fridman: pp.50/51, 60/61 (right), 62, 79

John Gerlach: pp.128, 137, 138, 139, 141,146/147 (left, top left, top right),

E. Hertzog: pp.134/135, 151,162/163, 164, 165, 166/167,

W.S. Keller: pp.35, 36/37 (bottom)

J. Kinsley: pp.168/169, 173

Museum of the American Indian: pp.42 (left), 88/89, 92 (right), 93 (bottom)

National Park Service: pp.58, 124/125 (top, bottom)

W.L. Rusho: pp.38/39, 50/51 (bottom), 60/61 (left), 64/65 (top right, bottom left), 81/82 (centre)

S.L. Convention and Visitors Bureau: pp.42 (bottom), 43 (left, right, bottom), 44/45, 49, 64/65 (top left, centre), 66/67, 68/69 (left, right), 70/71, 72, 73, 86, 87, 129, 130/131, 132, 133, 136, 149

US Department of the Interior: pp.24/25 (left), 56/57 (centre)

Trevor Wood: pp.9, 11, 15 (left), 26/27, 47, 52/53, 54/55, 56/57 (top, right), 59, 83, 84/85 (top),95, 97, 100/101, 102/103 (centre, top), 105, 106/107, 110/111, 114, 116, 120, 122, 123, 124/125 (centre, top), 140, 142/143, 144/145, 146/147 (bottom right), 151, 154, 156/157, 158/159, 160/161, 170/171, 172